# Current Topics in Microbiology 245/II and Immunology

Editors

Springer
Berlin
Heidelberg
New York
Barcelona
Hong Kong
London
Milan
Paris
Singapore
Tokyo

# Signal Transduction and the Coordination of B Lymphocyte Development and Function II

## Translation of BCR Signals to Specific Physiologic Outcomes

Edited by L.B. Justement and K.A. Siminovitch

With 13 Figures and 3 Tables

Springer

Louis B. Justement, Ph.D.
Associate Professor
Division of Developmental & Clinical Immunology
Department of Microbiology
University of Alabama at Birmingham
378 Wallace Tumor Institute
Birmingham, AL 35492-3300
USA
*e-mail*: Louis.Justement@ccc.uab.edu

Professor Katherine A. Siminovitch, M.D.
Department of Medicine
University of Toronto
Mount Sinai Hospital
600 University Ave., Rm. 656A
Toronto, Ontario
CANADA M5G 1X5
*e-mail*: ksimin@mshri.on.ca

*Cover Illustration: Part II – Schematic representation of signal transduction through the Pre-B cell antigen receptor (BCR) complex and the mature BCR complex expressed on either immature or mature B cells. Signal transduction via the Pre-BCR or mature BCR mediates distinct biological outcomes depending on the developmental/differentiation state of the B cell. The effect of cross-linking the Pre-BCR is designated by red arrows, the effect of cross-linking the mature BCR on immature B cells is designated by purple arrows and the response to cross-linking the BCR on mature B cells is designated by green arrows.*

*Cover Design: design & production* GmbH, Heidelberg

ISSN 0070-217X

ISBN-13: 978-3-642-64064-3 e-ISBN-13: 978-3-642-59641-4
DOI: 10.1007/978-3-642-59641-4


Softcover reprint of the hardcover 1st edition 2000

Library of Congress Catalog Card Number 15-12910



Typesetting: Scientific Publishing Services (P) Ltd, Madras

Production Editor: Angélique Gcouta

SPIN: 10717170 27/3020 – 5 4 3 2 1 0 – Printed on acid-free paper

# Preface

Proper development and differentiation of B lymphocytes is essential to ensure that an organism has the ability to mount an effective humoral immune response against foreign antigens. The immune system must maintain a balance between the deletion of harmful self-reactive B cells and the generation of a diverse repertoire of B cells that has the ability to recognize an almost unlimited array of foreign antigens. The need to delete self-reactive cells is tempered by the need to avoid the generation of large functional holes in the repertoire of foreign antigen-specific B cells that patrol the periphery. To accomplish this, the immune system must reach a compromise by eliminating only the most dangerous autoreactive clones, while allowing less harmful autoreactive B cells to exist in the periphery where they may complement the organism's ability to mount a rapid response against invading micro-organisms. Those autoreactive cells that do enter the peripheral pool are subject to a number of conditional restraints that effectively attenuate their ability to respond to self-antigens. Deleterious alterations in the homeostasis between tolerance induction and recruitment of B cells into the functional repertoire may lead to increased susceptibility to autoimmune disease or infection, respectively. Therefore, delineation of the molecular processes that maintain immunological homeostasis in the B cell compartment is critical.

The balance between tolerance and immunity can be altered by factors that are extrinsic as well as intrinsic to the B cell leading either to anergization and elimination, or to activation and differentiation into antibody-secreting plasma cells or memory cells. Extrinsic factors that regulate the balance between tolerance and immunity include differences in the amount, avidity and timing of antigen presentation, and the expression of molecules by other immune cells or tissues that play a role in elimination, retention or expansion of B cells. Equally important is the regulation of the intrinsic signaling threshold that is required to trigger an antigen-specific B cell clone and which determines whether that cell undergoes tolerance induction versus activation in response to a given antigenic challenge. Regulation of signaling thresholds in individual B cell clones is affected by alterations in the expression or function of molecules that adjust the strength of the signal delivered via the BCR. Triggering thresholds can be determined by inherited genetic polymorphisms that affect the expression or activity of specific intracellular effector proteins, or they can be altered within an individual B cell clone by previous encounters with antigen that lead to anergization or memory cell formation.

Significant progress has been made towards delineation of the intrinsic molecular processes that regulate B lymphocyte immune function. Recent observations have provided a clearer picture of the interactive signaling pathways that emanate from the mature B cell antigen receptor (BCR) complex and the different precursor complexes that are expressed during development. Studies have also revealed that the net functional response to a given antigenic challenge is affected by the combined action of BCR-dependent signaling pathways, as well as those originating from various coreceptors expressed by B cells. The chapters in this volume provide a summary of current findings relating to the molecular control of B cell development and differentiation. Because it is virtually impossible to include a discussion of every aspect related to signal transduction in B lymphocytes, an effort has been made to focus on signaling through the BCR complex and co-receptors that regulate BCR-dependent signaling. Part 1 of the volume deals with the biochemical/molecular aspects of BCR-dependent signal transduction beginning with membrane proximal events and culminating with regulation of gene transcription in the nucleus. Part 1 also covers the molecular function of specific coreceptors that are involved in regulation of BCR signaling.

It is now well established that reversible tyrosine phosphorylation plays an important role in regulating B cell biology. In particular, binding of antigen to the BCR promotes the activation of several protein tyrosine kinases (PTK) that, in conjunction with protein tyrosine phosphatases (PTP), alter the homeostasis of reversible tyrosine phosphorylation in the resting B cell. The net effect is a transient increase in protein tyrosine phosphorylation that facilitates the phosphotyrosine-dependent formation of effector protein complexes, promotes targeting of effector proteins to specific microenvironments within the B cell and initiates the catalytic activation of downstream effector proteins. The role of protein tyrosine kinases and phosphatases in initiation and propagation of signal transduction via the BCR is discussed in the first chapter. Initiation of B lymphocyte activation is dependent on the tyrosine phosphorylation-dependent formation of multi-molecular effector protein complexes that activate downstream signaling pathways. The formation of such complexes was initially hypothesized to occur primarily via effector protein binding to the BCR complex itself. However, recent studies have demonstrated that productive signaling via the BCR is, in fact, dependent on tyrosine phosphorylation of one or more adapter proteins that play a crucial role in recruitment and organization of effector proteins at the plasma membrane. The second chapter provides a discussion of the SLP65/BLNK adapter protein and its role in recruitment and activation of key signal transducing effector proteins in the B cell. After the BCR has been engaged by antigen and the activation response has been initiated, numerous second messengers and intermediate signal transducing proteins are activated. These include the production of lipid second messengers by phosphatidylinositol 3-kinase, the hydrolysis of phosphatidylinositol 4,5-bisphosphate to yield diacylglycerol and 1,4,5-inositoltrisphosphate, and the mobilization of $Ca^{2+}$. Numerous intermediate signaling proteins are also activated, including the Ras and Rap1 GTPases, as well as the Erk, JNK and p38 MAP kinases. The third chapter provides a discussion of the current understanding of how these second messengers and intermediate signaling proteins function in a concerted manner to regulate transcription factor activation and gene transcription in the B cell. In the fourth chapter the role of the cytoskeletal apparatus in B cell activation is discussed. Studies from numerous cell systems indicate that the cytoskeletal apparatus not only provides structural integrity for a cell, but it also provides a cell with the capacity to compartmentalize and redistribute proteins in a dynamic manner in order to modulate signal transduction and, thus, cellular function. It is now becoming clear that the cytoskeletal apparatus plays an important role in B cell activation through its physical/functional interactions with signal transducing effector proteins. Finally, the ability of B lymphocytes to respond to antigen is regulated by the expression of specific genes that play a role in the activation response. The fifth chapter presents a discussion on Pax-5/BSAP, which is a DNA-binding protein that plays a pivotal role in controlling the expression of genes that are required for the B cell response to antigen. A number of potential targets for Pax-5/BSAP have been identified that are directly associated with the BCR, or with signal transduction in B cells. The current understanding of how Pax-5/BSAP function is regulated and its role in gene transcription is covered there.

It is now clear that signal transduction through the BCR can be modulated by a number of coreceptors, in effect to maximize the immune response when the B cell encounters a foreign antigen in the appropriate microenvironment, to attenuate the response in instances where the B cell response to a self-antigen might be detrimental to the host and, finally, to modulate B cell activation under conditions in which an adequate immune response has already been made. In general, it is apparent that B cell coreceptors function as docking structures that recruit specific effector proteins in a phosphotyrosine-dependent manner. Depending of the context in which a B cell encounters antigen, one or more coreceptors can be engaged and, based on their ability to recruit selected effector proteins, either enhance or attenuate signaling through the BCR complex. The final two chapters of Part 1 provide an overview of the CD19, CD22 and FcγRIIb coreceptors describing their molecular function as well as their role in regulation of BCR-dependent signal transduction and B cell immune function.

In Part 2 of this volume the role of BCR- and accessory molecule-dependent signal transduction in regulation of specific physiological processes associated with B cell development and differentiation will be covered. The BCR complex performs an essential function during B cell development and differentiation. In pro-B and pre-B cells, expression of the membrane immunoglobulin heavy chain of the BCR in conjunction with pseudolight chain gene products is essential for establishing allelic exclusion and for regulation of ordered gene recombination at the heavy and light chain loci. Subsequent expression of a mature BCR complex on the surface of immature B cells is important for tolerance induction and selection into the peripheral B cell pool. During the process of tolerance induction, binding of self-antigen to the BCR on B cells can lead to anergization, clonal elimination or receptor editing, depending on the qualitative/quantitative nature of the signal delivered and the developmental state of the B cell. Finally, once B cells have been selected and migrate to the periphery, the BCR complex mediates antigen-dependent activation and selection into the memory B cell pool. In the first chapter of Part 2, the discussion focuses on the well-documented differences in the response of immature and mature stage B cells to antigen. Based on current information it remains controversial as to whether the differential response of these populations reflects processes that are intrinsic to B cells at specific stages of development or, rather, reflect extrinsic processes that determine the functional outcome of antigen-dependent signaling in immature versus mature B cells. In this context, the molecular and cellular characteristics of immature B cells are discussed and an evaluation of whether they support a role for intrinsic versus extrinsic factors in regulating the outcome of BCR signaling presented.

Before the mature BCR complex is expressed on the surface of the B cell, the gene elements that encode the heavy and light chain polypeptides that comprise the receptor must be successfully recombined in a process termed V(D)J recombination. This recombination process is highly regulated during development with respect to the expression of recombinase activity and the susceptibility of specific loci to undergo rearrangement. The high degree of control imposed on V(D)J recombination, which is crucial for successful enforcement of allelic exclusion and for maintenance of the fidelity of rearrangement, is exerted at several levels including: (1) transcription of recombinase genes; (2) accumulation of recombinase gene products; and (3) accesibility of specific loci to recombinase activity. The second chapter summarizes what is currently known about the signaling pathways in the B cell that underlie each of the regulatory mechanisms above. Because V(D)J recombination is an error-prone process in terms of the actual joining of gene elements, it is subject to quality control at many levels, inherently allowing for multiple attempts at rearrangement and correction of the final recombined gene product. Additionally, immune tolerance to self-antigen can be established in emerging B cells by reactivating the capability to undergo V(D)J recombination. Alternatively, self-tolerance can be effectively imposed by triggering B cells to undergo apoptosis. The third chapter focuses on what is currently known about the ability of the BCR to mediate both enhancing and suppressing effects on B cell development, V(D)J recombination and apoptosis.

Once B cells have encountered foreign antigen in the appropriate peripheral microenvironment, numerous soluble and cell-associated factors play a crucial role in driving proliferation and differentiation of the antigen-stimulated cell. The fourth chapter discusses current knowledge pertaining to signal transduction through CD40, which has been shown to play a critical role in regulating B cell growth and differentiation, class switching, germinal center formation and the generation of memory cells. Once B cells have encountered antigen and have received additional signals through CD40, they mature as antigen-presenting cells resulting from the upregulation of CD23, B7.1, B7.2 ICAM1 and CD44. In the fifth chapter evidence is presented indicating that the antigen-processing function of B cells is regulated both developmentally and in response to an array of stimuli that are received by the B cell through cell surface signaling molecules, including the BCR, the CD21/CD19 complex, MHC class II, CD40 and FcγRIIb. Finally, as the B cell response to antigen progresses, it is accompanied by class switching in which the isotype of antibody being synthesized is changed. This process does not affect the inherent specificity for antigen, but it does alter the effector function of the antibody produced. Class switching has been shown to occur in response to signals delivered to the B cell through cytokine receptors, CD40 and the BCR complex itself. The sixth chapter discusses what is currently known about the specificity of isotype switch recombination with regard to the regulation of germline transcripts by signals delivered through cell surface receptors.

The topics covered in this two-part volume are intended to provide information on the biochemical/molecular aspects of signal transduction through the BCR and to relate the mechanistic processes associated with signal transduction in B cells to immunologically relevant functional outcomes. It is also intended as a general overview of the molecular processes that underlie B cell development and differentiation and hopefully will serve as a reference for those who wish to delve further into the issues discussed herein or related topics.

L.B. JUSTEMENT
K.A. SIMINOVITCH

# List of Contents

# List of Contents of Companion Volume 245/I

# List of Contributors

(Their addresses can be found at the beginning of their respective chapters.)

# B-Cell Antigen Receptor Signaling in Immature-Stage B Cells: Integrating Intrinsic and Extrinsic Signals

J.G. Monroe



## 1 Introduction

The consequences and mechanism of B-cell antigen receptor (BCR) signaling in mature immunocompetent B cells are areas that have been well-studied and are beginning to be understood at the molecular level. This signaling determines whether a B cell with a particular antigen specificity and affinity will be recruited into an immune response. Of equal importance to the immune system is the process of BCR signaling in immature-stage B cells. Here, signaling through the BCR is associated with developmental arrest (Hartley et al. 1993), abortive cell-cycle progression (Carman et al. 1996), and cell death (Norvell et al. 1995) of the antigen-reactive B cell. The consequence of these responses to the organism is immune tolerance to secondary exposure to the particular antigen.

The differential response of immature and mature-stage B cells to antigens has been well documented [see review (Klinman 1996)]. It remains controversial as to whether these differences reflect processes that are intrinsic to B cells at these different stages of development or instead reflect extrinsic processes that determine

Department of Pathology and Laboratory Medicine, University of Pennsylvania School of Medicine, Room 311 BRBII/III, 421 Curie Blvd., Philadelphia, PA 19104, USA
e-mail: monroej@mail.med.upenn.edu

the outcome of antigen-induced primary signaling in these two populations of B cells. In this chapter, I will discuss molecular and cellular characteristics of immature B cells and evaluate to what degree they support a role for intrinsic versus extrinsic influences on the fate of these cells on BCR signaling.

## 2 B-Lymphocyte Development

In mature mammals, B cells are derived from multipotential progenitor cells localized to the endosteum of the bone marrow (JACOBSEN and OSMOND 1990). Commitment to the B-lymphocyte lineage is manifest during the pro-B-cell stage (Fig. 1). From there, maturation to the mature B cell occurs in a stepwise fashion (HARDY et al. 1991; OSMOND et al. 1998), regulated by intrinsic programs and extrinsic factors such as cytokines and cognate interactions with cellular components of the bone marrow and possibly the spleen (reviewed in JARVIS and LEBIEN 1998; KINCADE et al. 1998; MONTECINO-RODRIQUEZ and DORSHKIND 1998).

The early pro- and pre-B stages of B-cell development are characterized by ordered rearrangement of the genes encoding the immunoglobulin (Ig) heavy and light chains that will form the antigen-recognition component of the BCR. The Igα and Igβ signaling components of this receptor are expressed as early as the pro-B cell, before heavy- or light-chain proteins are expressed (ROLINK and MELCHERS 1991; ROLINK and MELCHERS 1993). More and more evidence suggests that Igα and Igβ play important roles in the developmental progression and expansion of early developing B cells. For example, B-cell development in mice deficient in Igβ expression is arrested at the transition between V-DJ and VDJ rearrangement (fraction B → C, GONG and NUSSENZWEIG 1996), implicating Igβ signaling in transition through the pro-B stage, before heavy-chain rearrangement is even complete. Mice that cannot express the pre-B form of the BCR because they do not express the surrogate light-chain protein λ5 (Kitamura et al. 1992) or lack the ability to express the membrane-associated form of the μ-heavy chain (KITAMURA et al. 1991) show severe developmental defects at the pro- to pre-B cell transition (fraction C → D). These observations and others (NAGATA et al. 1997) have led to the evolving idea that alternative forms of the BCR play critical roles in the developmental progression of B cells prior to expression of the mature or conventional BCR.

The complete BCR complex consisting of the Igα/Igβ signaling component and the tetrameric IgH/IgL antigen-recognition component is first expressed as developing B cells transit from the pre-B compartment into the immature (fraction D → E). Expression of the complete BCR on the surface of developing B cells defines the transition into the immature stage. Phenotypically, pre-B cells are nearly identical to the earliest immature B cells. Although levels of expression of some surface-expressed proteins such as CD22, B220 (CD45), human serum albumin (HSA), and class II proteins differ quantitatively (Fig. 2), these differences exist within a continuum where expression levels progressively change as B cells progress from the pre-B to the mature B-cell stage. It is not known whether pre- and

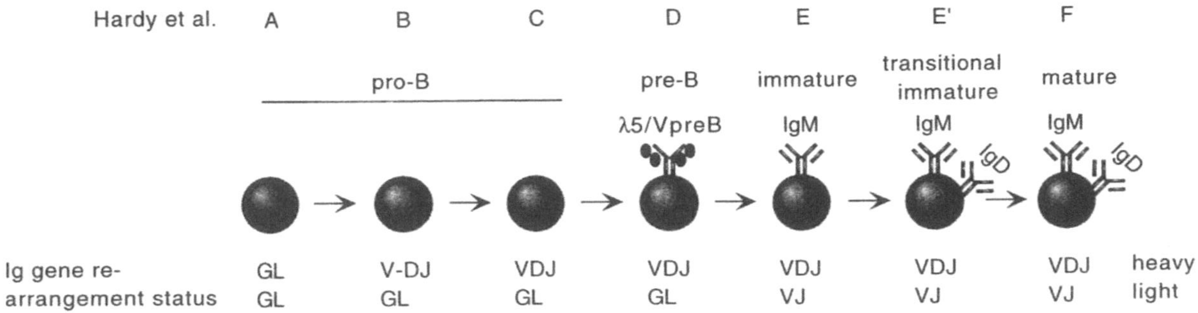


**Fig. 1.** Linear staging of B-cell development in the bone marrow

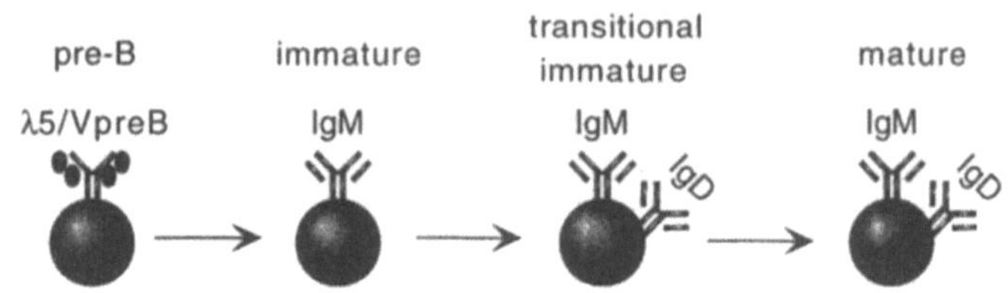


| | pre-B | immature | transitional immature | mature |
|---|---|---|---|---|
| HSA | +++ | ++++ | +++ | + |
| CD45(B220) | + | ++ | +++ | ++++ |
| CD19 | ++ | ++ | +++ | ++++ |
| IgM | +/- | ++++ | +++ | ++ |
| IgD | - | - | ++ | ++++ |
| CD22 | +/- | + | ++ | ++++ |
| CD23 | - | - | -/+++ | ++++ |

**Fig. 2.** Phenotypic characterization of B cells during the pre- to mature B transitions

immature-stage B cells manifest functional or physiological differences. However, the possibility that pre-B and immature B cells might respond differently to BCR signaling is an important point to consider as more focus is turned towards the study of immature B cells in Ig transgenic mice. In these models, Ig surface expression is no longer a marker for pre- to immature B transition, as it is aberrantly expressed as early as the pro-B stage of development.

## 3 Immature-Stage B Cells

Immature B cells first arise in the bone marrow of adult mammals. The bone-marrow immature B cells express the IgM form of the BCR exclusively. It is, therefore, at this stage that the B cell can first interact with the constellation of self-antigens expressed by the animal. Consequently, as will be discussed later, this is the first point where negative selection against self-reactive specificities occurs.

Until recently, it was believed that emigration from the bone-marrow microenvironment coincided with transition from the immature to mature stage. As a consequence, it has long been difficult to understand how negative selection to antigens expressed outside of the bone marrow (peripheral antigens) can occur. To accommodate negative selection to peripheral antigens, a number of selection processes have been proposed to allow mature B cells to distinguish between self and non-self antigens (GOODNOW et al. 1995). Although each of these probably play an important role in regulating responses by self-reactive mature-stage B cells, it is likely that negative selection to peripheral antigens by developing B cells is much more simple. Studies by ALLMAN et al. (1992, 1993) have definitively shown that the most recent emigrants from the bone marrow retain many of the phenotypic characteristics of the bone-marrow immature B cells. These transitional cells express high levels of HSA, which is the principal phenotypic difference between immature and mature B cells (Fig. 2). Most importantly, they remain highly sensitive to tolerance induction (ALLMAN et al. 1992) and respond to BCR signaling in vitro by undergoing apoptosis (NORVELL and MONROE 1996; SATER et al. 1998).

Recently, we have found the transitional immature B-cell population which has emigrated to the spleen to be comprised of two distinct subsets: CD23– and CD23+ populations (Fig. 3). Temporal studies using sublethal irradiated autoreconstituting mice show that repopulation of the periphery by $CD23^-/HSA^{hi}$ transitional cells precedes that of $CD23^+/HSA^{hi}$ cells, suggesting that the $CD23^-$ cells are precursors of the $CD23^+$ cells. Both populations respond to BCR crosslinking in vitro by apoptosis, so the functional relevance of these distinct transitional subsets is presently unknown. Of potential importance is the observation that levels of the orphan chemokine receptor Burkitt's lymphoma receptor 1 (BLR1) are different in these two subpopulations of transitional immature B cells. BLR1 expression in mature B cells is associated with homing into splenic follicles (FORSTER et al. 1996; SCHMIDT et al. 1998). On splenic immature B cells, BLR1 is expressed in over 98% of the $CD23^+$ transitional immature subset but only in 50–70% of the $CD23^-$ subset (Sater and Monroe, unpublished observations), suggesting that these two transitional immature subsets may be directed to distinct splenic microenvironments based on their chemokine receptor expression. This interpretation is consistent with studies of Lortran et al. (LORTRAN et al. 1987), who observed that recent bone-marrow emigrants do not migrate directly into the follicles but rather, are localized first to the red pulp and T cell zones.

Although the bone-marrow immature B cells are IgD–, expression of this isotype at low levels appears to be coincident with emigration into the periphery. Both the $CD23^-$ and $CD23^+$ peripheral subsets express detectable levels of IgD; expression shows a gradual increase as the maturation proceeds through the $CD23^-/HSA^{hi}$ (immature) to $CD23^+HSA^{hi}$ (transitional immature) to $CD23^+HSA^{lo}$ (mature) transitions. These results indicate that despite previous

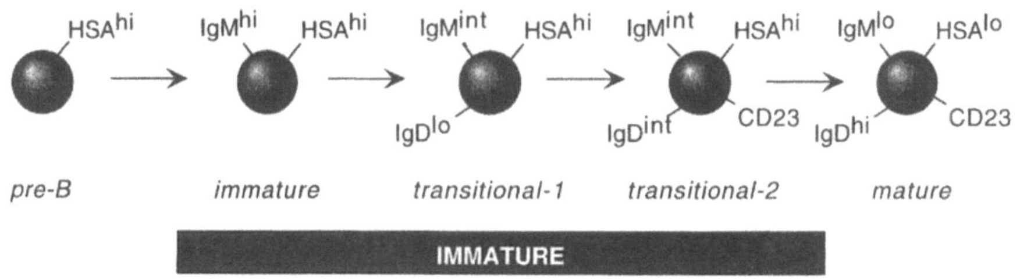


**Fig. 3.** Distinguishing phenotypes of immature-stage B cells

models of B-cell development, IgD expression does not distinguish immature-stage B cells from mature B cells but rather is a continuum as cells transit from immature to transitional and finally to mature stages (Figs. 2, 3). Furthermore, although it was once thought that IgD expression was coincident or even causal in the shift in functional responsiveness of immature and mature B cells to BCR engagement, there is now considerable evidence that IgD expression neither marks nor confers positive responsiveness to the B cell (as discussed later).

## 3.1 Functional Responses of Immature B Cells

As I have discussed, the prime immunological relevance of the immature B-cell stage is to provide a window for negative selection against B cells with self-reactive BCR. The idea that developing B cells are uniquely predisposed to negative responses to antigen was suggested by experiments dating back more than 50 years (reviewed in MONROE et al. 1993). In vitro (CAMBIER et al. 1976; METCALF and KLINMAN 1976; PIKE et al. 1980) and later in vivo (NEMAZEE and BURKI 1989; HARTLEY et al. 1993) studies have clearly established that encounter with antigen at this stage of B-cell development leads ultimately to elimination of the antigen-reactive clone. However, very little is known regarding the cellular and molecular processes that apparently direct immature-stage B cells towards negative selection. Presumably, these processes would define characteristics that distinguish immature B cells from their mature B-cell counterparts that are positively selected or activated by BCR signaling.

In vitro studies by our laboratory using purified immature-stage B cells from bone marrow (NORVELL et al. 1995), neonatal spleen (CARMAN et al. 1996), or adult-mouse spleen (NORVELL and MONROE 1996; SATER et al. 1998) have established that mature and immature B cells respond very differently to BCR signaling. Whereas mature B cells enter the cell cycle and proceed through at least one division (SOLVASON et al. 1996), immature B cells are signaled within 30min to enter a program that culminates in apoptosis (SATER et al. 1998). This apoptotic response is associated with abortive cell-cycle entry that results in a mid-G1 block (CARMAN et al. 1996). Bone marrow and neonatal splenic immature B cells upregulate the cell-cycle regulator cyclinD2 and proceed to activate the cyclin-dependent kinase CDK4, processes that are indicative of early G1 entry. However, they fail to upregulate the late G1 cyclin cyclinE, or form cyclinE/cdk2 complexes and proceed into S phase. Agents, such as interleukin (IL)-4, that block the apoptotic response (NORVELL et al. 1995; SATER et al. 1998) fail to bypass this block, indicating that the arrest in mid-G1 phase is not a consequence of cell death. The initial pathways leading to apoptosis appear to be reversible (SATER et al. 1998), but are soon (within 8–10h) followed by the induction of an apparently irreversible caspase-mediated apoptotic pathway (Sandel and Monroe, submitted). The differential response of the immature and mature B cell to BCR signaling predicts differences in the nature of the signals generated and/or the way in which these signals are processed or translated.

## 3.2 Structure of the Immature B-Cell BCR

Immature and mature B cells can differ markedly in their relative expression of the IgD and IgM forms of their BCR. As discussed above, immature B cells ($IgM^{hi}$ $IgD^{neg}$) progressively acquire more IgD as they develop into mature B cells ($IgM^{lo}$ $IgD^{hi}$). Since this progression is associated with a shift from a tolerance-sensitive phenotype to a tolerance-insensitive one, it has been argued that signaling through the IgD form of the BCR results in a positive signal for B-cell activation whereas the IgM form generates negative signals. Using this argument, immature B cells that are either $IgD^{neg}$ or $^{lo}$ would encounter antigen predominately via the IgM form. As B cells proceed further in their transition to mature B cells, BCR signaling would be increasingly dominated by signals generated through IgD. Here the negative signaling through IgM is postulated to be overridden by the positive signals generated through the IgD form of the BCR.

A differential role for IgM and IgD in generating signals for tolerance versus activation is supported by reports that have suggested that the two isotypes differ in their association with potential signal-transduction proteins (Kim et al. 1994; Terashima et al. 1994). There are also functional studies that support differential signaling via IgM versus IgD. For example, modulation or enzymatic removal of IgD from the surface of mature B cells was shown to make these cells more sensitive to tolerance induction (Cambier et al. 1977; Scott et al. 1977; Vitetta et al. 1977). In each of these cases, the ability to activate the B cell and its resistance to tolerance induction was directly associated with expression of IgD. Other in vitro studies using lymphoma models have been inconclusive regarding this question. Although some studies have suggested that IgM generates an exclusively negative signal (Ales-Martinez et al. 1988; Tisch et al. 1988), others have failed to establish that IgM and IgD generate signals that result in different responses by the B cell (Webb et al. 1989).

Several more recent studies have been designed to determine the importance of IgD in differentiating between signals for negative selection or activation. Using hen-egg lysozyme (HEL)-binding Ig transgenic mice whose B cells express only IgM or IgD or both, Brink et al. (1992) showed that IgM and IgD are indistinguishable in their ability to induce deletion, anergy, or proliferation. These results were consistent with another genetic mouse model in which it was shown that IgD-deficient mice are capable of responding to both T-dependent and T-independent antigens (Nitschke et al. 1993). Together, these studies indicate that the presence of IgD is not required for either the prevention of tolerance or the induction of activation responses by mature B cells.

The majority of studies argue that neither IgM nor IgD is responsible for determining the outcome of BCR signaling and instead suggest that it is the developmental stage that determines the response of the B cell, rather than the specific isotype engaged. Norvell and Monroe (1996) directly compared the relative contribution of IgM/IgD expression versus developmental stage in determining the response of B cells to BCR engagement. Using IgM/IgD co-expressing transitional immature and mature B cells, monoclonal anti-$\mu$ or anti-$\delta$ antibodies were used to signal selectively through IgM or IgD. In these studies, the ability of either isotype to trigger activation or negative selection was directly determined by using proliferation or apoptosis as in vitro readouts. The results clearly showed that the stage of development dictated the response of the B cell. In these studies, proliferation was triggered only in the mature B cells regardless of the isotype engaged. More importantly, either isotype triggered an apoptotic response in the transitional immature B cells. Together, these latter studies argue that the ratio of IgD to IgM expression plays no role in regulating the differential responsiveness of immature and mature B cells. This interpretation does not rule out that the form of the BCR could affect the ability of the B cell to engage particular types of antigens and thereby affect quantitative aspects of BCR signaling. In this regard, the longer hinge region associated with the IgD isotype (Nisonoff 1982) might confer increased ability to prolong engagement with antigens expressing fixed and/or highly spaced epitopes.

## 3.3 Co-Receptors

It has been consistently noted that immature B cells are extremely sensitive to tolerance induction, negative selection, and cell death. This sensitivity is evident in that it takes much lower concentrations of ligand or weaker receptor/ligand affinities to trigger these responses in immature B cells than it does to trigger positive responses in mature B cells (Sater et al. 1998; Lang et al. 1996).

The high sensitivity of the immature B-cell BCR makes teleological sense. It predicts that negative selection at the clonal level would occur at antigen concentrations and affinity thresholds lower than those necessary to trigger a later activation response by a mature B cell expressing the identical heavy/light-chain combination. As such, negative selection is dominant over activation and thereby ensures a high probability of eliminating clones with a potential for self-reactivity.

The factors that regulate the threshold at which BCR signals trigger responses by immature B cells are unclear. Since immature B cells express two- to fivefold higher levels of IgM than mature B cells, one might conclude that the sensitivity of these cells to BCR-induced responses reflects the level of antigen receptor expressed. However, when one considers the total BCR expression (IgM plus IgD), the levels of expression at these two stages of development are comparable (Sater et al. 1998). Nonetheless, alterations in other positive as well as negative modulators of BCR-signal strength could affect the sensitivity of the receptor and thereby determine the threshold of ligand engagement necessary to trigger a response by the immature B cell.

It is now appreciated that BCR signaling is modulated by a handful of co-receptor molecules that serve to increase or decrease the threshold of engagement necessary to initiate a BCR signal. The CD19/CD21 complex appears to be involved in enhancing the signal generated through the BCR, whereas CD22 expression is associated with decreasing the strength of the signal, thereby raising the threshold of antigen–receptor interaction needed to initiate a response by the B cell (Carter and Fearon 1992; Doody et al. 1995). Each of these co-receptor molecules is expressed on all immature B cell subsets as well as on mature B cells. However, as mentioned previously, their expression levels change as B-cell maturation proceeds through the immature-to-mature B-cell window (Fig. 2). Therefore, the exact balance in expression of positively and negatively modulating co-receptors could result in different strengths of signal to the same antigen encountered by the B cells as they transit from the immature to mature stages of development.

CD22 was originally proposed to augment BCR-induced proliferative signaling (Tedder et al. 1997), but a negative modulatory role for CD22 was suggested by experiments in which CD22 sequestration away from the BCR complex augmented BCR-induced responses (Doody et al. 1995). The cytoplasmic domain of CD22 contains tyrosine-based motifs capable of binding both positive and negative regulators of BCR-signal transduction (Campbell and Klinman 1995; Doody et al. 1995; Lankester et al. 1995; Law et al. 1996; Tuscano et al. 1996). The existence of both types of regulatory motifs on the CD22 cytoplasmic domain suggests that its effects on BCR signaling are complex. The negative regulatory motifs (immunoreceptor tyrosine-based inhibition motifs) have been shown to facilitate association with the tyrosine phosphatase *src*-homology-region-2 phosphatase (SHP)-1 (Campbell and Klinman 1995; Doody et al. 1995; Lankester et al. 1995; Law et al. 1996). SHP-1 is thought to be a negative regulator of BCR-induced signaling (Pani et al. 1995). Indeed, its loss of function in motheaten (me) and motheaten viable ($me^v$) mice (Shultz et al. 1993; Tsui et al. 1993) are associated with B-cell hyper-responsiveness and increased serum Ig.

CD22 is not expressed until the pre-B cell stage (Erickson et al. 1996; Nitschke et al. 1997). Consequently, early B-cell development in CD22-deficient mice progresses normally. However, mature B cells appear hyper-responsive to receptor signaling in that BCR-induced $Ca^{2+}$ fluxes are increased and achieved at lower levels of BCR cross-linking (O'Keefe et al. 1996; Otipoby et al. 1996; Sato et al. 1996a; Nitschke et al. 1997). The apparent chronic activation of $CD22^{-/-}$ B cells, coupled with the similarity in phenotype to B cells from SHP-1-deficient $me^v$/$me^v$ mice, suggests that CD22 negatively regulates BCR signaling, perhaps via binding of SHP-1. In the absence of CD22, the BCR may be spontaneously activated or the activation threshold may be lowered sufficiently so that it is stimulated following exposure to endogenous antigens.

As previously discussed, CD22 expression increases continuously as B cells progress from the pre-B stage to the mature B cell stage. The reduced level of CD22 expression in immature B cells relative to mature B cells and the consequential reduced ability to recruit SHP-1 to the vicinity of the BCR might predict that the antigen receptor of the former may be more sensitive to antigen encounter. Consistent with this idea is the observation that elimination of self-reactive immature B cells occurs at lower receptor/epitope avidity in SHP1-deficient mice than when SHP1 is present and presumably associated with CD22 (Cyster and Goodnow 1995). Thus, it appears that the developmentally regulated expression of CD22 may account or at least contribute to their sensitivity to BCR-induced negative selection.

The story for co-receptor CD19 is somewhat more complicated to interpret with regard to its involvement in regulating immature B cell responses. CD19 expression is one of the earliest markers of commitment to the B lineage. Expressed first on fraction-B pro-B cells, CD19 expression gradually increases during the transition to mature B cells (Sato et al. 1996b). The CD19 co-receptor has been hypothesized to play a positive role in regulating BCR-induced activation signals. In this regard, co-cross-linking of CD19 and the BCR results in an enhanced response to anti-Ig (Carter and Fearon 1992) and overexpression of CD19 is associated with elevated serum antibody levels (Engel et al. 1995). Mice that lack CD19 exhibit decreased serum antibody levels, impaired proliferative responses to anti-Ig (Engel et al. 1995; Sato et al. 1995), and marked impairment of B-cell responses to thymus-dependent antigens (Engel et al. 1995; Rickert et al. 1995). These characteristics exist despite normal B-cell development in the bone marrow (Engel et al. 1995; Rickert et al. 1995).

The ability of CD19 to enhance BCR-initiated responses may be due to its phosphotyrosine-mediated association with a number of signal-transduction proteins. Association of CD19 with inositol-3 kinase (PI-3 kinase), proto-oncogene Vav, and protein tyrosine kinases Lyn, Lck, and Fyn (Tuveson et al. 1993; Uckun et al. 1993; van Noesel et al. 1993b; Weng et al. 1994; Chalupny et al. 1995) have been reported. Recruitment of these signaling molecules to the BCR could additively influence ongoing BCR signaling by augmenting BCR-linked pathways, or even synergistically by initiating complementary, but unique, signaling pathways.

In addition to its effects on BCR signaling in mature B cells, CD19 has the ability to influence the sensitivity of BCR signaling in immature B cells as well. Studies in CD19-overexpressing transgenic mice suggest that an increase in CD19 expression augments the sensitivity of the immature B cell for negative selection (Engel et al. 1995). In these mice, there is a significant decrease in the size of their immature B-cell compartment. The decrease in the number of immature B cells could be due to fewer precursors entering the immature B-cell pool and/or a decrease in the transit time spent at the immature stage. However, the size of the pre-B-cell compartment appears to be unaffected by over-expression of CD19 while the mature compartment is decreased. These data may suggest that higher frequencies of immature B cells are being negatively selected in these mice, though this remains to be established. As a consequence, there are fewer surviving immature B cells to serve as precursors for the mature B-cell compartment. Operationally, receptors with affinities for endogenous antigen too low to stimulate negative selection in normal mice may be eliminated when CD19 is overexpressed.

In immature and mature B cells, CD19 is complexed with the complement receptor CD21 and TAPA-1/Leu-13. It is through co-ligation with the complement receptor CD21 that CD19 is believed to be brought into the BCR-signaling complex (van Noesel et al. 1993a; Fearon and Carter 1995; Carroll 1998). This interaction at least in part may explain the similar defects in adaptive immune responses observed in mice deficient in C3 or C4, Cr2 (CD21) and CD19 (Engel et al. 1995; Rickert et al. 1995; Ahearn et al. 1996; Croix et al. 1996; Fischer et al. 1996).

CD21, through its interaction with CD19, may also function to enhance negative selection of immature B cells with receptors directed towards endogenous antigens. In this regard, CD21$^{-/-}$*lpr*/*lpr* mice exhibit increased levels of anti-double-stranded DNA and rheumatoid-factor antibodies than do *lpr*/*lpr* mice with wild-type CD21 expression (Goerg et al. 1998). While these results may imply more relaxed negative selection of developing B cells with these specificities, it remains to be established whether or not their increased levels are due solely to impaired peripheral tolerance induction of mature-stage self-reactive B cells. Indeed, the size of the immature B-cell compartment in CD19$^{-/-}$ mice is not increased (Engel et al. 1995), suggesting that CD19 is not absolutely necessary for deletion of auto-reactive B cells.

Although the levels of CD19 expressed by immature B cells may be low enough that further loss does not affect negative selection in general, its presence in the immature B cell may serve a function in enhancing negative selection to a specific class of endogenous antigens. It has been postulated that some natural antibodies have the ability to recognize and bind a subset of self-antigens (Carroll 1998). This interaction would be predicted to activate the classical pathway of complement activation and result in the coating of the antigen with complement protein C3d. This self-antigen/C3d complex would allow co-ligation of the CD21/CD19 complex with the BCR. This co-ligation would lower the signaling threshold for the immature B-cell BCR, resulting in more efficient negative selection.

While the over-expression studies discussed above suggest that CD19 can influence the sensitivity of the BCR on immature B cells and potentially lower the signaling threshold through the BCR required in order to trigger negative selection, the CD19-deficient mouse studies suggest that its role under normal circumstances may be marginal. However, under normal circumstances CD19 expression on immature B cells may be sufficiently low so that the further loss in the CD19$^{-/-}$ animals may have little consequence on negative selection. Nonetheless, future studies may establish that what expression there is may in some cases enhance the ability of immature B cells to respond to specific self-antigens.

Like that of CD19, CD45 expression in mature B cells appears to act as a positive regulator of B-cell activation. Its expression in B-cell lines is critical for p21$^{ras}$ activation (Kawauchi et al. 1994; Pao et al. 1997), $Ca^{2+}$ mobilization (Justement et al. 1991), and recruitment of the protein tyrosine kinase Lyn into the BCR complex (Pao and Cambier 1997). However, unlike CD19, which appears to enhance the strength of BCR signals, CD45 probably acts to facilitate signal initiation by the BCR. Analysis of signal transduction events in B cells from CD45-deficient mice revealed that, although Ig$\alpha$ and phospholipase C$\gamma$2 are phosphorylated following BCR cross-linking, there appears to be an abrogation in the influx of calcium from extracellular stores (Benatar et al. 1996).

B cells from CD45-deficient mice have been suggested to resemble late immature or transitional B cells from non-transgenic mice because they exhibit a reduced ability to proliferate in response to anti-BCR antibodies (Kong et al. 1995; Benatar et al. 1996; Blyth et al. 1996). However, in contrast to traditional immature B cells, they do not undergo apoptosis (Kong 1995) nor do they express surface markers consistent with this stage of development (Sater and Monroe, unpublished observations).

CD45 has recently been suggested to play a role in setting the threshold between negative and positive selection of immature B cells into the long-lived B-cell pool (Cyster et al. 1996). Somewhat surprisingly, exposure of CD45-deficient HEL-binding immature B cells to soluble HEL in vivo led to an increased recovery of IgD+ B cells. These results suggest that in the absence of CD45, a normally tolerogenic signal is converted to a positively selecting one, and further suggest that a BCR-induced response of very low affinity may be required for efficient entry of B cells into the long-lived pool.

At this point, it remains unclear what precise effect the different levels of CD45 expression in immature and mature B cells have on quantitative and qualitative aspects of their BCR signaling. For the most part, the differences in expression levels are minor, reflecting two- to fourfold differences between early immature and transitional B cells and subsequent mature B cells (Sater and Monroe, unpublished).

The influence of relative expression of each of these co-receptor molecules in determining the apparent enhanced sensitivity of the immature B-cell BCR to antigen is complicated by altered levels of both positive as well as negative modulators of BCR signaling. Decreased expression of CD22 in immature B cells relative to that in mature B cells could easily account for the enhanced sensitivity of the BCR in the immature B-cell populations. However, the simultaneous decreases in expression of positive (CD45, CD19) modulators complicates this simplistic interpretation. Importantly, the relative levels of expression of signaling-effector molecules that associate with these co-receptors in immature and mature B cells has not been thoroughly studied. Unpublished studies (Sillman and Monroe, unpublished observations) have indicated that transitional and bone-marrow immature B cells express SHP-1 at levels approximating those in mature B cells. However, the complexity of the interplay between co-receptor expression and that of the associated signaling proteins will make it difficult to assess their individual roles with regard to BCR signaling during the immature-to-mature transition by conventional knockout and overexpression transgenic approaches. More useful may be hemizygous genetic approaches, such as those used by Cornall et al. (Cornall et al. 1998) to analyze the role of protein tyrosine kinase Lyn, tyrosine phosphatase SHP-1, and co-receptor CD22 in BCR signaling of mature B cells. This approach, although still not ideal, affords some ability to analyze complex relationships between multiple signaling molecules that converge on a single cellular response.

## 3.4 Intrinsic Versus Extrinsic Determination of the Fate of Immature-Stage B Cells to BCR Signaling

Differences in the ratio of IgM to IgD, coupled with subtle but real differences in co-receptor expression likely play a role in determining the types of antigens that will be most efficient in triggering immature B cells. Furthermore, co-receptor levels may be important in regulating quantitative aspects of BCR triggering in the immature B cell. In particular, the relatively low expression of CD22 likely contributes to, and may in fact account for, the enhanced sensitivity of the antigen receptor on these cells.

Despite their influence on the quantitative aspects of BCR signaling at these stages of development, none of the differences described above would appear to account for the qualitatively different responses that are elicited in immature and mature B cells by BCR cross-linking. In this regard, the molecular processes that control the fate decision leading to either negative selection or activation are for the most part completely undefined. In a more general sense, it is not completely clear that the differences observed in the qualitative response of immature and mature B cells to antigens reflect intrinsic differences in the B cells. It remains a matter of controversy as to whether there exist intrinsic or "hardwired" differences between immature and mature B cells that determines their default response to BCR signaling. Alternatively, the response of the immature B cell to antigen-receptor engagement may be determined entirely by B-cell extrinsic factors such as availability of T-cell help; BCR signals in this case would be identical with those of mature B cells. Of course, these scenarios are not mutually exclusive, and I will argue that both intrinsic and extrinsic factors influence the ultimate fate of the immature B cell on BCR signaling.

## 3.5 Extrinsic Determination of Negative Selection of Immature B Cells

According to the original model of Bretscher and Cohn, B lymphocytes stimulated through their antigen receptor (signal 1) in the absence of T-cell help (signal 2) would be tolerized (Bretscher and Cohn 1968). The relative lack of T cells in the bone marrow that could provide a "second signal" to immature B cells exposed to self antigen raises the possibility that negative selection of immature B cells occurs because of a lack of T-cell help. This paradigm remains the framework for the dual signal models of both T- and B-cell activation and tolerance induction. It is likely that this is a mechanism for the selective inactivation and elimination of mature B and T cells. In both instances, initiation of an activation program without secondary co-stimulatory signals leads to aborted activation, anergy, and eventually death (Hartley et al. 1993). However, in order for such a mechanism to solely regulate immature B-cell negative selection, one would have to predict that in the absence of T-cell-mediated co-stimulatory signals, both immature and mature B cells should respond identically to BCR signaling. More specifically, in the absence of T-cell-derived secondary signals, both immature and mature B cells should undergo negative selection following BCR signaling. However, this does not appear to be the case. For example, despite a lack of T-cell help, highly purified populations of mature B cells proliferate in response to anti-Ig stimulation in vitro. In contrast, immature B cells undergo apoptosis under identical conditions (Norvell et al. 1995; Norvell and Monroe 1996). Thus, these results suggest that an intrinsic, developmentally regulated difference in BCR-induced signal transduction is responsible, at least in part, for the functional dichotomy between antigen-stimulated immature and mature B cells.

While the experiments described above suggest that B-cell tolerance is not a direct consequence of inadequate T-cell help, T-cell help can have a profound influence on the induction of immature B-cell tolerance (Metcalf and Klinman 1976, 1977). In particular, ligation of the B-cell surface antigen CD40 by CD40L expressed on activated T cells, or the $T_h$ lymphokine IL-4, can efficiently rescue immature B cells from anti-Ig-induced apoptosis in vitro (Chang et al. 1991; Brines and Klaus 1992; Sater et al. 1998), and activated $T_h$ cells can prevent tolerance induction in vivo (Fulcher et al. 1996).

In conclusion, I would argue that B-cell intrinsic factors define the developmental differences in the response of immature and mature B cells to BCR signaling. These factors link the BCR in B cells of each of these developmental stages to distinct default responses. It is important to note that although these intrinsic factors trigger default response programs in the B cell, these responses can nonetheless be modified by signals extrinsic to the B cell. Before further discussing the immunological relevance of these extrinsic signals I will discuss what is currently understood regarding the intrinsic processes that distinguish immature and mature B cells during BCR signal transduction and signal processing.

## 3.6 BCR-Relevant Intrinsic Factors that Distinguish Immature and Mature B Cells

All of the studies already discussed have implied a developmentally associated distinction between BCR signaling in immature and mature B cells. That this difference is intrinsic to the B cell and not a consequence of the influence of extrinsic factors, such as access to T-cell help (discussed above), is strongly supported by the fact that, following their isolation from other cells, immature and mature B cells respond differently to identical BCR cross-linking (Norvell et al. 1995; Norvell and Monroe 1996). What then remains to be established, and what I will discuss now, is the molecular basis for the intrinsic differences between immature and mature B cells that account for their profound differential responses to BCR engagement. We and others have identified a number of functional and molecular differences between these two developmentally distinct B-cell populations. In general, there exist differences that would affect both the qualitative and quantitative aspects of the BCR-induced signal. The quantitative influences I have already discussed in the context of co-receptor expression. I would now like to turn my attention to molecular characteristics of the immature and mature B cells that could determine the fate of the cell after BCR cross-linking. In Fig. 4, I have diagrammed three potential models, for which there is some experimental support, that could explain how signaling through the BCR could lead to very different outcomes in immature and mature B cells.

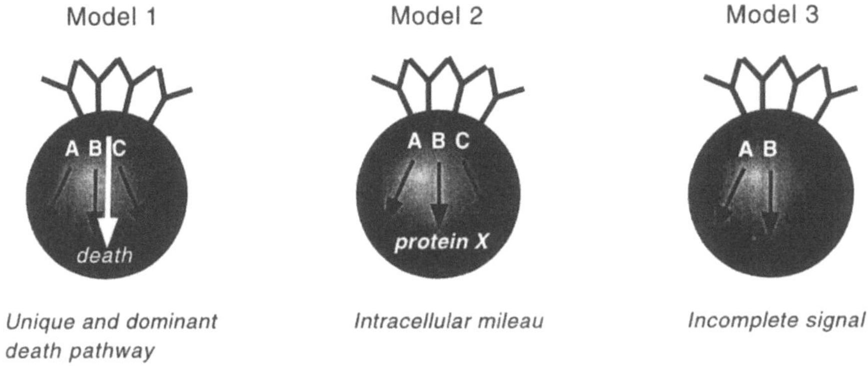


**Fig. 4.** Potential models to account for the differential responses of immature and mature B cells to B-cell antigen receptor signaling

The first model proposes that the immature B-cell BCR is "hardwired" to a death pathway that is not initiated by signaling in mature B cells. In immature B cells, this pathway overrides any of the positive signaling pathways that might be shared with BCR signal transduction in mature B cells. Support of this model could be inferred from studies of GILBERT et al. (1993), which showed that the intracellular enzyme phospholipase A2 is expressed only in immature B cells and thymocytes. They suggest that activation of phospholipase A2 may be involved in signaling of immature B-cell apoptosis because it is activated in the WEHI-231 "immature" B lymphoma only under conditions where apoptosis is triggered.

Formal arguments to prove or disprove the existence of a specific BCR-coupled death pathway in immature B cell do not yet exist. However, there are several observations that would appear to argue that if it does exist it cannot account for all aspects of the differential responses associated with BCR signaling in immature and mature B cells. For example, IL-4 (CHANG et al. 1991; NORVELL et al. 1995; SATER et al. 1998) or specific peptide inhibitors of caspase-3 (Sandel and Monroe, submitted) can block the apoptotic response of the immature B cell to BCR cross-linking. However, in our hands, the B cells still do not proceed through the cell cycle as they would following BCR cross-linking of mature B cells. These results would seem to indicate that the absence of an activation response by immature-stage B cells to BCR cross-linking is not due to the fact that they die before such a response can occur. Moreover, they establish that even in the absence of cell death BCR signaling in immature and mature B cells is different.

The second model used to explain developmentally associated differences in the response of immature and mature B cells to BCR signaling involves differences in the intracellular milieu of the cells at these distinct stages of development. The scenario in this case is that BCR signal initiation and second-messenger generation in immature and mature B cells are identical. However, there exist developmentally associated differences in expression of an intracellular protein, "protein X", that defines the intracellular milieu such that identical signaling pathways result in different outcomes. Logical candidates for this protein(s) are members of the *BCL-2* gene family. The balance of anti- and pro-apoptotic members of this family has been shown to exert a profound influence on cell survival in a number of systems.

The prototypic member of the *BCL-2* extended gene family is *bcl-2*. *Bcl-2* encodes a 25- to 26-kDa protein predominantly located in the membranes of the mitochondria, endoplasmic reticulum, and nuclear envelope (Hockenbery et al. 1990). The Bcl-2 protein forms either homodimers or heterodimers with an ever-expanding range of *bcl-2* family members including, but not limited to, pro-apoptotic members such as *bax* (Oltvai et al. 1993) and *bad* (Yang et al. 1995), as well as anti-apoptotic molecules such as *bcl-x*$_L$ (Boise et al. 1993) and *A1* (Lin et al. 1993). It has been theorized that the relative expression of these family members functions as a rheostat mechanism, setting the threshold for determining whether a cell lives or dies when exposed to potentially lethal stimuli (Oltvai and Korsmeyer 1994). What remains to be fully established is to what degree members of the *bcl-2* family directly determine whether a signal will lead directly to survival or apoptosis, or will instead play a more secondary role in determining the sensitivity of the cell to apoptotic stimuli. In the latter case, the relative expression of anti-apoptotic bcl-2 proteins would determine the level of resistance of the cell to an apoptotic signal.

The expression patterns of *bcl-2*, *bcl-x*$_L$, and *A1* have been evaluated throughout B-cell development and exhibit differential regulation. *Bcl-2* is most highly expressed in pro- and mature B cells, while expression in pre-B and immature B cells is low (Li et al. 1993; Merino et al. 1994). *Bcl-x*$_L$, however, exhibits its highest expression in pre-B cells, with some expression in pro- and immature-stage cells (Grillot et al. 1996). Expression of *bcl-x*$_L$ in mature B cells is nearly absent, although it can be induced by a variety of mature B cell stimuli, including BCR cross-linking (Choi et al. 1996; Grillot et al. 1996). *A1* expression is low in immature and transitional immature B cells, but its mRNA expression increases by tenfold as mature B cells transit into the long-lived pool (Tomayko and Cancro 1998).

Expression of these *bcl-2* family members appears to affect B cells uniquely at different stages of B-cell development. For example, B-cell development in mice deficient for *bcl-2* expression is normal until 3–4 weeks of age, at which time massive lymphoid apoptosis occurs (Nakayama et al. 1993; Vies et al. 1993). This apoptosis occurs in the peripheral lymphoid organs and therefore reflects a requirement for *bcl-2* in the maintenance of the peripheral B-cell pool. Consistent with this hypothesis, *bcl-2*-overexpressing mice do have increased numbers of peripheral mature B cells, presumably because of the increased life span of these cells (Strasser et al. 1991). Although *bcl-2*-overexpressing mice exhibit autoimmune symptoms late in life, it is not clear whether this break in tolerance is because of a defect in central tolerance or because peripheral tolerance is compromised by a reduced ability to delete autoreactive mature B cells. In contrast, the absence of *bcl-x*$_L$ appears to result in reduced survival of pre-B cells (Motoyama et al. 1995), suggesting that upregulation of *bcl-x*$_L$ by engagement of the pre-BCR may play an important role in B-cell development. Over-expression of *bcl-x*$_L$ in the B-cell compartment of transgenic mice results in an accumulation of pre-B, immature and mature B cells as a consequence of increased survival of pre-B cells rather than decreased negative selection of immature B cells (see below) (GRILLOT et al. 1996). In contrast, mice deficient in the pro-apoptotic family member *bax* displayed an increased number of B cells with a mature B-cell phenotype, suggesting that a proportion of B cells may be eliminated through a *bax*-dependent mechanism (KNUDSON et al. 1995), perhaps during negative selection.

The ability of *bcl-2* and *bcl-$x_L$* to regulate negative selection of immature B cells has been tested directly, although it is difficult to formulate a model that is consistent with all of the available data. Studies by LANG et al. (1997) argue that forced expression of *bcl-2* prevents apoptosis that results when an immature B cell encounters antigen. In their system, apparently large frequencies of self-reactive immature B cells are able to undergo secondary light-chain gene re-arrangement and replace their self-reactive receptors with non-reactive ones. In the absence of ectopic *bcl-2* expression, B cells that fail to accomplish this receptor-editing process successfully die, either because they fail to express any receptor or because the subsequent receptor(s) maintain self-reactivity. As expected, prevention of receptor editing by crossing these mice onto a recombinase-activating gene (*RAG*)-deficient background results in profound depletion of peripheral B cells. Interestingly, forced expression of *bcl-2* in these mice leads to accumulation of IgM−, B220+ cells in the periphery, suggesting either that negative selection by cell death is not occurring or, alternatively, death due to the absence of positive selection is prevented.

Conflicting results were observed by NISITANI et al. (1993) for bone-marrow B cells reactive to erythrocyte antigen. Again, there was an increase in the number of IgM−, B220+ cells in the spleens of *bcl-2* transgenic mice with B cells expressing self-reactive receptors. However, in this case, the overall frequency of B220+ cells in the spleen was markedly decreased relative to that observed in the antigen-non-reactive mice, suggesting that deletion of self-reactive B cells was occurring at a high frequency despite forced expression of *bcl-2*.

Other studies also support the argument that *bcl-2* expression is ineffective in blocking deletion of self-reactive immature B cells. HARTLEY et al. (1993) crossed mice transgenic for anti-HEL immunoglobulin onto *bcl-2*-expressing transgenic mice and then adoptively transferred bone marrow from these mice into membrane HEL (mHEL)-expressing or non-expressing recipients. In the absence of transgenic *bcl-2*, there was a decrease in B220+ cells in the periphery from 56% to 6% in the presence of self-antigen. Although the decrease in the transgenic *bcl-2*-expressing B cells was less, it was still high; from 81% B220+ cells in the absence of antigen down to 40% in the presence of antigen. Moreover, the remaining B220+ cells were arrested at the immature B-cell stage and, although they exited from the bone marrow, they continued to maintain a relatively short half-life and died before transiting into the long-lived mature B-cell pool. These results suggest that although forced expression of *bcl-2* can in some cases delay deletion of self-reactive immature B cells, it does not allow their continued development into mature cells. As a consequence, they die within several days due to their intrinsic short lifespan.

Considering that *bcl-$x_L$* and not *bcl-2* expression might be more effective in modulating negative selection of immature B cells, FANG et al. (1998) used the anti-HEL-immunoglobulin transgenic model to test the influence of forced *bcl-$x_L$*

expression on deletion of self-reactive immature B cells. Again, bone marrow from HEL-reactive immunoglobulin transgenic mice was adoptively transferred into mHEL-expressing or non-expressing mice. In B cells that expressed *bcl-x*$_L$, there was a dramatic inhibition of deletion in the presence of mHEL. In addition, some (38%) of the surviving self-reactive B cells showed evidence for receptor editing, again supporting the conclusion that receptor editing may be an alternative fate for self-reactive immature B cells when deletion is prevented or delayed. As a note, it is interesting that not all of the surviving B cells replaced their self-reactive receptors with non-antigen-reactive ones, or lost receptor expression entirely. This could indicate that the majority of self-reactive B cells never attempt to edit their receptors or, alternatively, that the majority of secondary light chains generated in this model cannot successfully compete with the transgenic light chain for heavy-chain pairing.

The differential expression of *bcl-2* and *A1* in immature and mature B cells would argue that these cells might differ in their inherent survival. In this regard, the turnover rate for immature B cells is on the order of 3–4 days, whereas it is over 8 weeks for mature B cells (Osmond 1986; Forster et al. 1989; Forster and Rajewski 1990; Allman et al. 1993). Moreover, although both immature and mature B cells express low levels of *bcl-x*$_L$, these levels increase after stimulation through the BCR in mature B cells. Definitive studies in immature B cells regarding the ability to induce *bcl-x*$_L$ have not been completed. We have observed a similar increase in *bcl-x*$_L$ mRNA levels in immature and mature B cells after anti-immunoglobulin-mediated BCR cross-linking (Sater and Monroe, in preparation). The results described above, although conflicting, would argue that *bcl-2* and/or *bcl-x*$_L$ expression could in some cases modulate BCR-induced deletion of immature B cells. The differences in its ability to do so in all of the cases discussed above could be related to a number of variables. Foremost could be the relative levels of *bcl-2* or *bcl-x*$_L$ expressed. Studies of Grillot et al. (1996) showed that *bcl-2*/*bcl-x*$_L$ double transgenic B cells have increased survival over either family member expressed alone, suggesting a dosage effect for survival. A related possibility is that the strength of the BCR signal may determine to what degree these anti-apoptotic proteins will be effective. If one considers that these proteins function as rheostats for regulating a B cell's inherent sensitivity to death, their normal absence in the immature B cell would cause these cells to be extremely sensitive to death signals that are triggered as a consequence of BCR engagement. Forced expression of these proteins could protect these cells from death and allow them to undergo fates such as anergy or receptor editing that normally would occur only as a consequence of weaker signals or in the presence of survival-promoting microenvironments when expression of these proteins is lower.

By the above argument, the differential expression of *bcl-2*/*bcl-x*$_L$/*A1* or other *bcl-2* family members may well contribute to the determination that immature B cells will be very sensitive to a signal for death once it is generated, but that expression levels of these proteins do not determine whether the BCR signal will be coupled to a negative selection or activation response. Support for this tentative conclusion comes from the observations above that, even when deletion is prevented or delayed by forced expression of these proteins, the immature B cells still do not undergo positive activation responses. Therefore, the differential response of immature and mature B cells to BCR engagement is likely determined by other aspects of BCR signaling.

The third model to account for differential responses to immature and mature B cells proposes that signaling is incomplete in the immature B cells. In this case, only a subset of the signaling pathways initiated in mature B cells is coupled to the BCR at the immature stage of development. In this model, receptor engagement leads to an attempted activation response that is ultimately aborted. As a consequence of this aborted activation response, the B cell perceives an incomplete signal and a default apoptotic pathway is initiated.

First noted in splenic B cells from neonatal mice, immature B cells are less efficient than mature B cells in their ability to hydrolyze inositol phospholipids in response to BCR cross-linking (Yellen et al. 1991). This signaling pathway is responsible for triggering the initial release of $Ca^{2+}$ from intracellular pools and activating the serine/threonine kinase, protein kinase C (PKC). It accomplishes this by the generation of two second-messenger molecules, inositol trisphosphate ($IP_3$), which triggers $Ca^{2+}$ release, and diacylglycerol (DAG), which regulates the activity of certain members of the PKC family of kinases. Although immature B cells in this study exhibited BCR-induced release of intracellular $Ca^{2+}$ equivalent to that observed for mature B cells, they lacked detectable generation of $IP_3$. Interestingly, subsequent studies showed that downstream events believed to be linked to PKC activation, such as *c-myc* induction, are impaired (Carman et al. 1996). The differential uncoupling to individual components of the same signaling pathway argues that the triggering thresholds for each are set at different levels. Several years ago, these results suggested to us that an unbalanced signal consisting of increased intracellular $Ca^{2+}$ levels in the absence of PKC activation may account for the negative response of immature B cells to BCR cross-linking. Consistent with this hypothesis, King et al. (submitted) have observed that phorbol diester-induced activation of PKC blocks the apoptotic response of immature B cells to BCR cross-linking. Interestingly, a specific activator of PKCβ is sufficient to block BCR-induced negative selection in these cells. Further support for the hypothesis that PKC (and perhaps PKCβ, specifically) plays a determining role in the fate decisions made by immature and mature B cells is the fact that phorbol diester-mediated depletion of DAG-regulated PKC isoforms in mature splenic B cells converts a BCR-induced activation response into an apoptotic one. King et al. (King et al. 1999) further showed that this treatment favors depletion of PKCα and PKCβ isoforms, again implicating PKCβ as the mediator of this fate decision.

Clearly, pharmacologic studies limit our ability to conclusively assign a role for PKCβ in determining these responses. A final test of this model will depend on genetic manipulation of PKCβ activity in B cells as well as further biochemical studies designed to define the mechanism by which PKCβ could manipulate these responses. With regard to the former, studies in mice that are deficient in PKCβ confirm that PKCβ plays an important role in the activation responses of mature B cells (Leitges et al. 1996). PKCβ-deficient mice respond poorly to anti-Ig antibodies in vitro, although it remains to be seen whether this unresponsiveness is due to an inadequate activation signal or to the induction of apoptosis, as would be predicted by the above model. Other PKC isoenzymes appear unable to compensate for the deficiency in PKCβ, suggesting that PKCβ is either uniquely regulated following BCR crosslinking or has distinct substrate specificity. Interestingly, the phenotype of PKCβ-deficient mice resembles that observed in the $Btk^{-/-}$ mouse (Kerner et al. 1995; Khan et al. 1995). Since the pleckstrin homology domain of Btk binds PKC (Yao et al. 1994), it is possible that these molecules are intimately associated during BCR-induced signal transduction and B-cell activation.

With regard to the biochemical aspects of how PKCβ might regulate these differential responses, there are clues; however, no firm conclusions can yet be made. As discussed above, immature and mature B cells differ in their induction of the *c-myc* proto-oncogene. While mature B cells upregulate *c-myc* as a consequence of BCR cross-linking (Snow et al. 1986), immature B cells do not (Carman et al. 1996). An inability to maintain *c-myc* expression has been shown to play a role in the BCR-induced apoptosis of WEHI-231 cells (reviewed in Sonenshein 1997). Treatment of WEHI-231 cells with agents that lead to an elevation in the NFκB inhibitor Iκβ resulted in a decrease in *c-myc* expression and the subsequent induction of cell death, an effect that could be reversed by the ectopic expression of *c-myc* (Wu et al. 1996a,b). The inability to upregulate *c-myc* expression following BCR engagement in immature B cells may play a pivotal role in their decision to undergo apoptosis. Interestingly, phorbol diesters are able to activate NFκβ (Baldwin 1996; Stancovski 1997) and have also been shown to upregulate *c-myc* transcription in immature B cells (Carman et al. 1996), suggesting a possible mechanism for its ability to block the BCR-induced apoptotic response in immature B cells. Again, whether the effects of phorbol diester are mediated through PKCβ remains to be determined.

## 4 Modulation of Negative Selection

Based on the studies discussed in the previous sections, I argue that the default pathway for BCR signaling (at least for high-avidity interactions) in immature B cells is one that leads to induced apoptosis and deletion. Moreover, properties intrinsic to the developmental stage of the B cell determine the default response to BCR cross-linking. While this can account for the observation that the immature B cell is the primary target for negative selection, it does not explain how negative selection can occur via multiple mechanisms (i.e. by processes other than apoptotic deletion).

Studies by several laboratories have shown that high-avidity interactions with antigen by immature B cells in the bone marrow do not necessarily lead to deletion but instead may result in editing of the BCR (Gay et al. 1993; Lang et al. 1996). During receptor editing, recombinase-activating genes are re-expressed and further light- and heavy-chain gene rearrangement occurs, although light-chain rearrangement appears to be more common (GAY et al. 1993; RADIC et al. 1993; CHEN et al. 1995; PRAK and WEIGERT 1995; RADIC and ZOUAI 1997). Light chains produced during the secondary rearrangement can replace the original light chain, potentially leading to a loss of self-reactivity. Thus, while it appears that receptor editing could play a role in maintaining self tolerance, it is difficult to reconcile the fact that a BCR-induced signaling event can lead to receptor editing in cells which have presumably also received a signal to undergo BCR-induced apoptosis. Our recent studies (SATER et al. 1998) have shown that the apoptotic response by transitional immature B cells is initiated within 30min after BCR engagement. Therefore, the window of opportunity for receptor editing would appear to be unrealistically narrow.

To resolve this paradox, MELAMED et al. (1998) have recently suggested that early immature B cells may be predisposed to receptor editing, whereas the more mature transitional cells default toward deletion. These studies utilized an in vitro model system, in which Ig-transgenic pre-B cells are expanded by culture with IL-7. Upon release from IL-7, the B cells continue maturation toward a more mature phenotype. Assuming that the $IgM^{lo}$ cells represent the earliest immature-stage cells and that the $IgM^{hi}$ represent transitional immature B cells, they observed upregulation of *RAG* mRNA and receptor editing in the early, immature but not the transitional B cells. Furthermore, BCR-induced apoptosis was only observed in the more mature population.

While the above studies suggest a developmental association for the differential coupling of BCR signaling to receptor editing and apoptosis, it remains to be established if this mechanism operates in non-Ig-transgenic B cells, where the distinction between pre-B and immature B cell is more unequivocally definable. This distinction may be critical if signaling through BCR-like receptors is different in pre-B cells, which are not easily distinguished from early immature B cells in Ig-transgenic B cells due to the precocious expression of surface Ig by the transgene.

As an alternative to a developmental model for determining the mechanism of negative selection, we have considered that the different microenvironments in which immature-stage B cells are localized may provide different B-cell extrinsic signals that determine the fate of the immature B cell to antigen encounter. As illustrated in Fig. 5, developing B cells with the response pattern of immature B cells exist within at least four major anatomic compartments: (1) the cellular environment of the bone marrow, (2) the central sinusoids of the marrow, in which they accumulate and then egress into the periphery (JACOBSEN and OSMOND 1990), (3) the circulation, and finally (4) the different microenvironments of the spleen. Given these very different microenvironments, it might be predicted that the intrinsic response of the immature B cell to BCR cross-linking would be differentially influenced by extrinsic factors inherent to these anatomically diverse locations. One might also infer such a possibility from the studies of Carsetti et al. (CARSETTI et al. 1995), in which TNP-reactive immature B cells are sensitive to deletion only in vitro, where they are isolated from bone-marrow stromal cells.

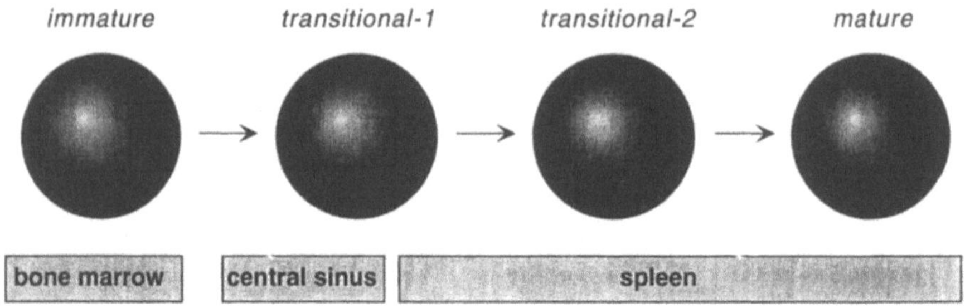


**Fig. 5.** Anatomic compartmentalization of immature-stage B cells throughout their transition toward the mature stage

The hypothesis that bone marrow provides an environment that allows alternative responses to BCR engagement was confirmed in recent studies (SANDEL and MONROE 1999). In this case, it was observed that while immature-stage B cells undergo apoptosis following BCR cross-linking, this apoptotic response is abrogated in the presence of unfractionated bone marrow. The protective effect is observed to be limited to this cellular environment, as neither whole spleen nor isolated, purified, mature B cells can confer this protection. Cell–cell contact appears to be required, as protection from apoptosis is not observed when the immature B cells are separated from the bone-marrow cells by a cell-impermeable membrane. These studies support the idea that a cellular component of adult bone marrow can alter the intrinsic response of the immature B cell to BCR cross-linking.

The cell(s) in the bone marrow responsible for the survival signals imparted to immature B cells and the molecular basis for this protection remain to be elucidated. Furthermore, the ultimate fate of immature B cells following antigen-receptor engagement in the context of this protective environment is not firmly established. With regard to the latter issue, in vitro cultures of highly purified transitional immature B cells on bone marrow cells derived from *RAG-2*-deficient mice exhibit a four- to sixfold induction of *RAG-2* mRNA expression. These results suggest that not only does the cellular environment of the bone marrow prevent BCR-induced apoptosis, it also provides signals that lead to recombinase re-induction. As a consequence, the intrinsic fate of the immature B cells on BCR engagement is altered to favor receptor editing rather than deletion.

Figure 6 outlines our interpretation of these studies with regard to the fate of the immature B cells upon encounter with antigens in the periphery and in bone-marrow compartments. Immature B cells encountering antigens in the presence of the protective marrow-derived cell (MDC) would be protected from antigen-induced cell death and would, therefore, have a window of opportunity in which to edit their receptors in response to antigenic stimulation (pathway 1). If receptors containing secondarily rearranged light chains also exhibit antigen reactivity, further rounds of receptor editing can occur as long as the immature B cell remains within the protective environment. Upon egress to the periphery, if there is no reactivity to antigenic epitopes present in this environment the immature B cell will

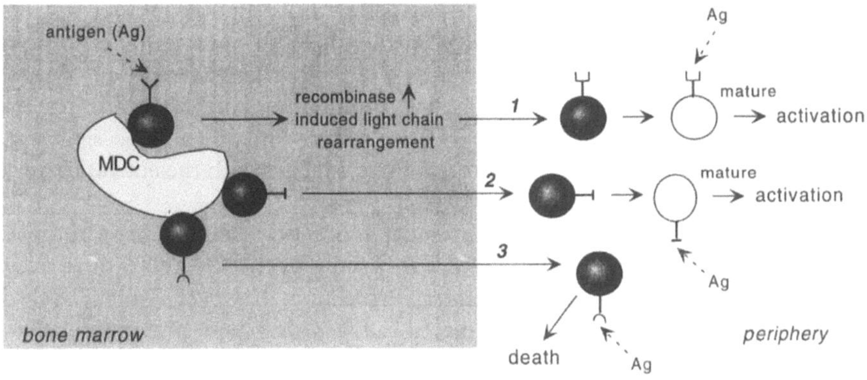


**Fig. 6.** Proposed fates of immature B cells with self- and non-self receptors in the bone marrow and periphery

further develop into a mature B cell that is capable of becoming activated in response to antigenic stimulation (pathway 1). Similarly, immature B cells non-reactive to antigens present in either the bone marrow or the periphery exit the marrow and transit into the mature B cell pool (pathway 2). In contrast, in the absence of the protective environment, immature B cells that encounter antigen in the periphery are not afforded the opportunity to edit their receptor and are eliminated by BCR-induced apoptosis (pathway 3). Likewise, immature B cells that have undergone receptor editing but which express BCRs that maintain self-reactivity, due either to failed receptor editing or to the expression of new receptors reactive to peripheral self antigens, will be induced to undergo apoptosis as a consequence of BCR signaling. It is unknown at this time whether the immature B cells leave the protective niche conferred by the MDCs and become sensitive to deletion while still in the marrow or instead do so only upon egress to the periphery. In either event, the critical aspect of this model is that there is an environmental signal that alters the fate of the immature B cell on BCR signaling.

Finally, I would like to consider the fate of immature B cells that enter the periphery with "non-self" receptors but then encounter foreign antigens before transiting into the mature, immunocompetant stage. The above discussion would imply that these B cells would be eliminated in light of the more general pressure to silence and remove autoreactive specificities. However, several lines of evidence argue that the fate of these cells can also be modified by processes that are rational within the context of the immune system.

Previous studies by us and others have suggested that signals derived from activated T cells may be able to block the apoptosis-inducing effects of BCR cross-linking on immature-stage B cells. For example, Scott et al. (1987) have shown, using the WEHI-231 cell line model of immature B cells, that anti-IgM induced growth inhibition and cell death is blocked by IL-4. More relevant are studies from our laboratory and others showing that neonatal spleen-derived immature-stage B cells can be protected from anti-Ig induced apoptosis by exogenous IL-4 or by a gp39 (CD40L)-expressing T-cell line (Chang et al. 1991; Brines and Klaus 1992; Norvell et al. 1995; Sater et al. 1998). Furthermore, this protection is conferred even if these agents are added to cultures as late as 8h after the addition of anti-Ig (Sater et al. 1998). After 8–10h of anti-Ig stimulation, the apoptotic response appears to be irreversible. The cut-off in the window in which these signals can modulate the BCR-induced apoptotic response correlates with the activation of the apoptotic executioner caspase-3 (Sater et al. 1998).

Together, these studies suggest that antigen-induced deletion can be abrogated in the periphery, provided that B cells can productively interact with activated T cells within 8h of antigenic encounter. Protection, however, would likely be contingent on the ability of the immature B cell to process and present antigens to the antigen-reactive T cell. It is not yet known whether peripheral immature B cells can present antigens to either resting or activated T cells in order to elicit the signals necessary for protection from apoptosis. It is also unclear whether immature B cells protected in this fashion could subsequently be activated to undergo clonal expansion and differentiation into antibody-secreting cells.

Thus, it would seem that the presence of non-B cells in different microenvironmental contexts could play a determining role in the fate of the immature B cell upon encounter with antigen. With appropriate regulation, these extrinsic signals can guide the immature B cell in the selection processes that underlie the self/non-self education process of the immune system.

*Acknowledgements.* I wish to thank Drs. Leslie King, Peter Sandel, Richard Sater, Amanda Norvell, Amy Yellen-Shaw, and Robert Wechsler, who have contributed to the work from my laboratory that is described in this review and who have been intellectual sources for development of the hypotheses that I have discussed.

## References

Ahearn J, Fischer M, Croix D, Goerg S, Ma M, Xia J, Zhou X, Howard R, Rothstein T, Carroll M (1996) Disruption of the Cr2 locus results in a reduction in B-la cells and in an impaired B cell response to T-dependent antigen. Immunity 4:251–262

Ales-Martinez JE, Warner GL, Scott DW (1988) Immunoglobulins D and M mediated signals that are qualitatively different in B cells with an immature phenotype. Proc Natl Acad Sci U S A 85:6919–6923

Allman DM, Ferguson SE, Cancro MP (1992) Peripheral B cell maturation. I. Immature peripheral B cells in adults are heat-stable antigen$^{hi}$ and exhibit unique signaling characteristics[1]. J Immunol 149:2533–2540

Allman DM, Ferguson SE, Lentz VM, Cancro MP (1993) Peripheral B cell maturation. II. Heat-stable antigen hi splenic B cells are an immature developmental intermediate in the production of long-lived marrow-derived B cells. J Immunol 151:4431–4444

Baldwin AS (1996) The NF-kB and IkB proteins: new discoveries and insights. Ann Rev Immunol 14:649–681

Benatar T, Carsetti R, Furlonger C, Kamalia N, Mak T, Paige CJ (1996) Immunoglobulin-mediated signal transduction in B cells from CD45-deficient mice. J Exp Med 183:329–334

Blyth KF, Conroy LA, Howlett S, Smith AJH, May J, Alexander DR, Holmes N (1996) CD45-null transgenic mice reveal a positive regulatory role for CD45 in early thymocyte development, in the selection of CD4+CD8+ thymocytes, and in B cell maturation. J Exp Med 183:1707–1718

Boise LH, Gonzalez-Garcia M, Postema CE, Ding L, Lindsten T, Turka LA, Mao X, Nunez G, Thompson CB (1993) Bcl-x, a bcl-2-related gene that functions as a dominant regualtor of apoptotic cell death. Cell 74:597–608
Bretscher PA, Cohn M (1968) Minimal model for the mechanism of antibody induction and paralysis by antigen. Nature 220:444
Brines RD, Klaus GGB (1992) Inhibition of lipopolysaccharide-induced activation of immature B cells by anti-μ and anti-δ antibodies and its modulation by interleukin-4. Int Immunol 4:765–771
Brink R, Goodnow CC, Crosbie J, Adams E, Eris J, Mason DY, Hartley SB, Basten A (1992) Immunoglobulin M and D antigen receptors are both capable of mediating B lymphocyte activation, deletion, or anergy after interaction with specific antigen. J Exp Med 176:991–1005
Cambier JC, Kettman JR, Vitetta ES, Uhr JW (1976) Differential susceptibility of neonatal and adult murine spleen cells to in vitro induction of B-cell tolerance. J Exp Med 144:293–297
Cambier JC, Vitetta ES, Kettman JR, Wetzel GM, Uhr JW (1977) B cell tolerance. III. Effect of papain mediated cleavage of cell surface IgD on tolerance susceptibility of murine B cells. J Exp Med 146:107–117
Campbell MA, Klinman NR (1995) Phosphotyrosine-dependent association between CD22 and protein tyrosine phosphatase 1 C. Eur J Immunol 25:1573–1579
Carman JA, Wechsler-Reya RJ, Monroe JG (1996) Immature stage B cells enter but do not progress beyond the early G1 phase of the cell cycle in response to antigen receptor signaling. J Immunol: in press
Carroll MC (1998) The role of complement and complement receptors in induction and regulation of immunity. Ann Rev Immunol 16:545–568
Carsetti R, Kohler G, Lamers MC (1995) Transitional B cells are the target of negative selection in the B cell compartment. J Exp Med 181:2129–2140
Carter RH, Fearon DT (1992) CD19: Lowering the threshold for antigen receptor stimulation of B lymphocytes. Science 256:105–107
Chalupny NJ, Aruffo A, Esselstyn JM, Chan PY, Bajorath J, Blake J, Gilliland LK, Ledbetter JA, Tepper MA (1995) Specific binding of Fyn and phosphatidylinositol 3-kinase to the B cell surface glycoprotein CD19 through their src homology 2 domains. Eur J Immunol 25:2978–2984
Chang T-L, Capraro G, Kleinman RE, Abbas AK (1991) Anergy in immature B lymphocytes. Differential responses to receptor-mediated stimulation and helper T lymphocytes. J Immunol 147:750–756
Chen C, Nagy Z, Prak EL, Weigert M (1995) Immunoglobulin heavy chain gene replacement: a mechanism of receptor editing. Immunity 3:747–755
Choi MS, Homan M, Atkins CJ, Klaus GGB (1996) Expression of bcl-x during mouse B cell differentiation and following activation by various stimuli. Eur J Immunol 26:676–682
Cornall RJ, Cyster JG, Hibbs ML, Dunn AR, Otipoby KL, Clark EA, Goodnow CC (1998) Polygenic autoimmune traits: Lyn, CD22, and SHP-1 are limiting elements of a biochemical pathway regulating BCR signaling and selection. Immunity 8:497–508
Croix D, Ahearn J, Rosengard A, Han S, Kelsoe G, Ma M, Carroll M (1996) Antibody response to T-dependent antigen requires B cell expression of complement receptors. J Exp Med 183:1857–1864
Cyster JG, Goodnow CC (1995) Protein tyrosine phosphatase 1C negatively regulates antigen receptor signaling in B lymphocytes and determines thresholds for negative selection. Immunity 2:13–24
Cyster JG, Healy JI, Kishihara K, Mak TW, Thomas ML, Goodnow CC (1996) Regualtion of B-lymphocyte negative and positive selection by tyrosine phosphatase CD45. Nature 381:325–328
Doody GM, Justement LB, Delibrias CC, Matthews RJ, et al. (1995) A role in B cell activation for CD22 and the protein tyrosine phosphatase SHP. Science 269:242–244
Engel P, Zhou LJ, Ord DC, Sato S, Koller B, Tedder TF (1995) Abnormal B lymphocyte development, activation, and differentiation in mice that lack or overexpress the CD19 signal transduction molecule. Immunity 3:39–50
Erickson LD, Tygrett LT, Bhatia SK, Grabsteins KH, Waldschmidt TJ (1996) Differential expression of CD22 (Lyb8) on murine B cells. Int Immunol 8:1121–1129
Fang W, Weintraub CB, B D, Garside P, Pape AK, Jenkins KM, Goodnow CC, Mueller LD, Behrens WT (1998) Self-reactive B lymphocytes overexpressing Bcl-$x_L$ escape negative selection and are tolerized by clonal anergy and receptor editing. Immunity 9:35–45
Fearon DT, Carter RH (1995) The CD19/CR2/TAPA-1 complex of B lymphocytes: linking natural to acquired immunity. [Review] Ann Rev Immunol 13:127–149
Fischer M, Ma M, Goerg S, Zhou X, Xia J, Finco O, Han S, Kelsoe G, Howard R, Rothstein T, Kremmer E, Rosen F, Carroll M (1996) Regulation of the B cell response to T-dependent antigens by the classical complement pathway. J Immunol 157:549–556

Forster I, Rajewski K (1990) The bulk of the peripheral B cell pool is stable and not renewed from the bone marrow. Proc Natl Acad Sci USA 87:4781–4784
Forster I, Vieira P, Rajewsky K (1989) Flow cytometric analysis of cell proliferation dynamics in the B cell compartment of the mouse. Internal Immunol 1:321–331
Forster R, Mattis AE, Kremmer E, Wolf E, Brem G, Lipp M (1996) A putative chemokine receptor, BLR1, directs B cell migration to defined lymphoid organs and specific anatomic compartments of the spleen. Cell 87:1–20
Fulcher DA, Lyons AB, Korn SL, Cook MC, Koleda C, Parish C, Fazekas de St Groth B, Basten A (1996) The fate of self-reactive B cells depends primarily on the degree of antigen receptor engagement and availability of T cell help. J Exp Med 183:2313–2328
Gay D, Saunders T, Camper S, Weigert M (1993) Receptor editing: an approach by autoreactive B cells to escape tolerance. J Exp Med 177:999–1008
Gilbert JJ, Wakelam MJ, Harnett MM (1993) A role for phospholipase D activation in B cell activation. Biochem Soc Trans 21:493
Goerg S, Chu L, Prodeus A, Zhou X, Carroll MC (1998) Deficiency in CD21 and CD35 increases autoimmune disease in lpr mice
Gong S, Nussenzweig MC (1996) Regulation of an early developmental checkpoint in the B cell pathway by Igβ. Science 272:411–414
Goodnow CC, Cyster JG, Hartley SB, Bell SE, Cooke MP, Healy JI, Akkaraju S, Rathmell JC, Pogue SL, Shokat KP (1995) Self-tolerance checkpoints in B lymphocyte development. Adv Immunol 59:279–368
Grillot DAM, Merino R, Pena JC, Fanslow WC, Finkelman FD, Thompson CB, Nunez G (1996) bcl-x exhibits regulated expression during B cell development and activation and modulates lymphocyte survival in transgenic mice. J Exp Med 183:381–391
Hardy RR, Carmack CE, Shinton SA, Kemp JD, Hayakawa K (1991) Resolution and characterization of pro-B and pre-pro-B cell stages in normal mouse bone marrow. J Exp Med 173:1213
Hartley SB, Cooke MP, Fulcher DA, Harris AW, Cory S, Basten A, Goodnow CC (1993) Elimination of self-reactive B lymphocytes proceeds in two stages: arrested development and cell death. Cell 72:325–335
Hockenbery D, Nunez G, Miliman C, Schreiber RD, Korsmeyer SJ (1990) Bcl-2 is an inner mitochondrial membrane protein that blocks programmed cell death. Nature 348:334–336
Jacobsen K, Osmond DG (1990) Microenvironmental organization and stromal cell associations of B lymphocyte precursor cells in mouse bone marrow. Eur J Immunol 20:2395–2404
Jarvis LJ, LeBien TW (1998) Cytokine and stromal influences on early B cell development. In: Monroe JG, Rothenberg EV (eds) Molecular biology of B-cell and T-cell development. Humana Press, Totowa, New Jersey, pp 231–251
Justement LB, Campbell KS, Chien N, Cambier JC (1991) Regulation of B cell antigen receptor signal transduction and phosphorylation by CD45. Science 252:1839–1842
Kawauchi K, Lazarus AH, Rapoport MJ, Harwood A, Cambier JC, Delovitch T (1994) Tyrosine kinase and CD45 tyrosine phosphatase activity mediate p21ras activation in B cells stimulated through the antigen receptor. J Immunol 152:3306–3316
Kerner JD, Appleby MW, Mohr RN, Chien S, Rawlings DJ, Maliszewski CR, Witte ON, Perlmutter RM (1995) Impaired expansion of mouse B cell progenitors in mice lacking Btk. Immunity 3:301–312
Khan WN, Alt FW, Gerstein RM, Malynn BA, Larsson I, Rathbun G, Davidson L, Muller S, Kantor AB, Herzenberg LA, Rosen FS, Sideras P (1995) Defective B cell development and function in Btk-deficient mice. Immunity 3:283–299
Kim K-M, Adachi T, Nielsen PJ, Terashima M, Lamers MC, Kohler G, Reth M (1994) Two new proteins preferentially associated with membrane immunoglobulin D. EMBO J 13:3793–3800
Kincade PW, Medina K, Smithson G, Zheng Z, Oritani K, Borghesi L, Yamashita Y, Payne K, Shimozato T (1998) Life/death decisions in B lymphocyte precursors: a role for cytokines, cell interaction molecules, and hormones. In: Monroe JG, Rothenberg EV (eds) Molecular biology of B-cell and T-cell development. Humana Press, Totowa, New Jersey, pp 177–196
King LB, Norvell A, Monroe JG (1999) Antigen receptor-induced signal transduction imbalances associated with the negative selection of immature B cells. J Immunol 162:2655–2662
Kitamura D, Kudo A, Schaal S, Muller W, Melchers F, Rajewsky K (1992) A critical role of lambda 5 protein in B cell development. Cell 69:823–831
Kitamura D, Roes J, Kuhn R, Rajewsky K (1991) A B cell-deficient mouse by targeted disruption of the membrane exon of the immunoglobulin μ chain gene. Nature 350:423–426
Klinman NR (1996) The "clonal selection hypothesis" and current concepts of B cell tolerance. Immunity 5:189–195

Knudson CM, Tung KS, Tourtellotte WG, Brown GA, Korsmeyer SJ (1995) Bax-deficient mice with lymphoid hyperplasia and male germ cell death. Science 270:96–99
Kong Y-Y, Kishihara K, Sumichika H, Nakamura T, Kaneko M, Nomoto K (1995) Differential requirements of CD45 for lymphocyte development and function. Eur J Immunol 25:3431–3436
Lang J, Jackson M, Teyton L, Brunmark A, Kane K, Nemazee D (1996) B cells are exquisitely sensitive to central tolerance and receptor editing induced by ultralow affinity, membrane-bound antigen. J Exp Med 184:1685–1697
Lankester AC, van Schijndel GM, van Lier RA (1995) Hematopoietic cell phosphatase is recruited to CD22 following B cell antigen receptor ligation. J Biol Chem 270:20305–20308
Law CL, Sidorenko SP, Chandran KA, Zhao A, Shen SH, Fischer EH, Clark EA (1996) CD22 associates with protein tyrosine phosphatase 1C, Syk, and phospholipase C-γ1 upon B cell activation. J Exp Med 183:547–560
Leitges M, Schmedt C, Guinamard R, Davoust J, Schall S, Stabel S, Tarakhovsky A (1996) Immunodeficiency in protein kinase Cβ-deficient mice. Science 273:788–791
Li Y-S, Hayakawa K, Hardy RR (1993) The regulated expression of B lineage associated genes during B cell differentiation in bone marrow and fetal liver. J Exp Med 178:951–960
Lin EY, Orlofsky A, Berger MS, Prystowski MB (1993) Characterization of A1, a novel hemopoietic-specific early-response gene with sequence similarity to bcl-2. J Immunol 151:1979–1988
Lortran JE, Roobottom CA, Oldfield S, MacLennan ICM (1987) Newly produced virgin B cells migrate to secondary lymphoid organs but their capacity to enter follicles is restricted. Eur J Immunol 17:1311–1316
Melamed D, Benschop RJ, Cambier JC, Nemazee D (1998) Developmental regulation of B lymphocyte immune tolerance compartmentalizes clonal selection from receptor selection. Cell 92:173–182
Merino R, Ding L, Veis DJ, Korsmeyer SJ, Nunez G (1994) Developmental regulation of the Bcl-2 protein and susceptibility to cell death in B lymphocytes. EMBO J 13:683–691
Metcalf ES, Klinman NR (1976) In vitro tolerance induction of neonatal murine B cells. J Exp Med 143:1327–1340
Metcalf ES, Klinman NR (1977) In vitro tolerance induction of bone marrow cells: a marker for B cell maturation. J Immunol 118:2111–2116
Monroe JG, Yellen-Shaw AJ, Seyfert VL (1993) Molecular basis for unresponsiveness and tolerance induction in immature stage B lymphocytes. Adv Mol Cell Immunol 1B:1–32
Montecino-Rodriquez E, Dorshkind K (1998) Regulation of lymphocyte development by microenvironmental and systemic factors. In: Monroe JG, Rothenberg EV (eds) Molecular biology of B-cell and T-cell development. Humana Press, Totowa, New Jersey, pp 197–212
Motoyama N, Wang F, Roth KA, Sawa H, Nakayama K, Nakayma K, Negishi I, Senju S, Zhang Q, Fujii S, Loh DY (1995) Massive cell death of immature hematopoietic cells and neurons in Bcl-x-deficient mice. Science 267:1506–1510
Nagata K, Nakamura T, Kitamura F, Kuramochi S, Taki S, Campbell KS, Karasuyama H (1997) The Igα/Igβ heterodimer on μ-negative proB cells is competent for transducing signals to induce early B cell differentiation. Immunity 7:559–570
Nakayama K, Negishi I, Kuida K, Shinkai Y, Louie MC, Fileds LE, Lucas PJ, Stewart V, Alt FW, Loh DY (1993) Disappearance of the lymphoid system in Bcl-2 homozygous mutant chimeric mice. Science:1584–1588
Nemazee DA, Burki K (1989) Clonal deletion of B lymphocytes in a transgenic mouse bearing anti-MHC class I antibody genes. Nature 337:562–566
Nisitani S, Tsubata T, Murakami M, Okamoto M, Honjo T (1993) The bcl-2 gene product inhibits clonal deletion of self-reactive B lymphocytes in the periphery but not in the bone marrow. J Exp Med 178:1247–1254
Nisonoff A (1982) In: Introduction of molecular immunology. Sinauer Associates, New York
Nitschke L, Carsetti R, Ocker B, Kohler G, Lamers MC (1997) CD22 is a negative regulator of B-cell receptor signalling. Curr Biol 7:133–143
Nitschke L, Kosco MH, Kohler G, Lamers MC (1993) Immunoglobulin D-deficient mice can mount normal immune responses to thymus-independent and -dependent antigens. Proc Natl Acad Sci USA 90:1887–1891
Norvell A, Mandik L, Monroe JG (1995) Engagement of the antigen-receptor on immature murine B lymphocytes results in death by apoptosis. J Immunol 154:4404–4413
Norvell A, Monroe JG (1996) Acquisition of surface IgD fails to protect from tolerance-induction. Both surface IgM- and surface IgD-medaited signals induce apoptosis of immature murine B lymphocytes. J Immunol 156: in press

O'Keefe TL, Williams GT, Davies SL, Neuberger MS (1996) Hyperresponsive B cells in CD22-deficient mice. Science 274:798–801
Oltvai ZN, Korsmeyer SJ (1994) Checkpoints of dueling dimers foil death wishes. Cell 79:189
Oltvai ZN, Milliman CL, Korsmeyer SJ (1993) Bcl-2 heterodimerizes in vivo with a conserved homolog, Bax, that accelerates programmed cell death. Cell 74:609–619
Osmond DG (1986) Population dynamics of bone marrow B lymphocytes. Immunol Rev 93:103–124
Osmond DG, Rolink A, Melchers F (1998) Murine B lymphopoiesis: towards a unified model. Immunol Today 19:65–68
Otipoby KL, Andersson KB, Draves KE, Klaus SJ, Farr AG, Kerner JD, Perlmutter RM, Law CL, Clark EA (1996) CD22 regulates tymus-independent responses and the lifespan of B cells. Nature 384:634–637
Pani G, Kozlowski M, Cambier JC, Mills GB, Siminovitch KA (1995) Identification of the tyrosine phosphatase PTP1C as a B cell antigen receptor-associated protein involved in the regulation of B cell signaling. J Exp Med 181:2077–2084
Pao LI, Bedzyk WD, Persin C, Cambier JC (1997) Molecular targets of CD45 in B cell antigen receptor signal transduction. J Immunol 158:1116–1124
Pao LI, Cambier JC (1997) Syk, but not Lyn, recruitment to B cell antigen receptor and activation following stimulation of CD45- B cells. J Immunol 158:2663–2669
Pike BL, Kay TW, Nossal GJV (1980) Relative sensitivity of fetal and newborn mice to induction of hapten-specific B cell tolerance. J Exp Med 152:1407–1412
Prak EL, Weigert M (1995) Light chain replacement: a new model for antibody gene rearrangement. J Exp Med 182:541–548
Radic MZ, Erikson J, Litwin S, Weigert M (1993) B lymphocytes may escape tolerance by revising their antigen receptors. J Exp Med 177:1165–1173
Radic MZ, Zouai M (1997) Receptor editing, immune diversification, and self-tolerance. Immunity 5:505–511
Rickert RC, Rajewsky K, Roes J (1995) Impairment of T-cell-dependent B-cell responses and B-1 cell development in CD19-deficient mice. Nature 376:352–355
Rolink A, Melchers F (1991) Molecular and cellular origins of B lymphocyte diversity. Cell 66:1081–1094
Rolink A, Melchers F (1993) B lymphopoiesis in the mouse. [Review] Adv Immunol 53:123–156
Sandel PC, Monroe JG (1999) Negative selection of immature B cells by receptor editing or deletion is determined by site of antigen encounter. Immunity 10:289–299
Sater RA, Sandel PC, Monroe JG (1998) BCR-induced apoptosis in primary transitional murine B cells: Signaling requirements and modulation by T cell help. Int Immunol 10:1673–1682
Sato S, Miller AS, Inaoki M, Bock CB, Jansen PJ, Tang ML, Tedder TF (1996a) CD22 is both a positive and negative regulator of B lymphocyte antigen receptor signal transduction: altered signaling in CD22-deficient mice. Immunity 5:551–562
Sato S, Ono N, Steeber DA, Pisetsky DS, Tedder TF (1996b) CD19 regulates B lymphocyte signaling thresholds critical for the development of B-1 lineage cells and autoimmunity. J Immunol 157: 4371–4378
Sato S, Steeber DA, Tedder TF (1995) The CD19 signal transduction molecule is a response regulator of B-lymphocyte differentiation. Proc Natl Acad Sci USA 92:11558–11562
Schmidt KN, Hsu CW, Griffin CT, Goodnow CC, Cyster JG (1998) Spontaneous follicular exclusion of SHP1-deficient B cells is conditional on the presence of competitor wild-type B cells. J Exp Med 187:929–937
Scott DW, Layton JE, Nossal GJV (1977) Role of IgD in the immune response and tolerance. I. Anti-δ pretreatment facilities tolerance induction in adult B cells in vitro. J Exp Med 146:1473–1483
Scott DW, O'Garra A, Warren D, Klaus GGB (1987) Lymphoma models for B cell activation and tolerance. VI. Reversal of anti-Ig mediated negative signalling by T cell-derived lymphokines. J Immunol 139:3924–3929
Shultz LD, Schweitzer PA, Rajan TV, Yi T, Ihle JN, Matthews RJ, Thomas ML, Beier DR (1993) Mutations at the murine motheaten locus are within the hematopoietic cell protein-tyrosine phosphatase (Hcph) gene. Cell 73:1445–1454
Snow EC, Fetherston JD, Zimmer S (1986) Induction of the c-myc protooncogene after antigen binding to hapten-specific B cells. J Exp Med 164:944–949
Solvason N, Wu WW, Kabra N, Wu X, Lees E, Howard MC (1996) Induction of cell cycle regulatory proteins in anti-immunoglobulin-stimulated mature B lymphocytes. J Exp Med 184:407–417
Sonenshein GE (1997) Down-modulation of c-myc expression induces apoptosis of B lymphocyte models of tolerance via clonal deletion. J Immunol 158:1994–1997

Stancovski IB, D (1997) NF-κB Activation: The IκB kinase revealed? Cell 91:299–302

Strasser A, Whittingham S, Vaux DL, Bath ML, Adams JM, Cory S, Harris A (1991) Enforced BCL2 expession in B-lymphoid cells prolongs antibody responses and elicits autoimmune disease. Proc Natl Acad Sci USA 88:8661–8665

Tedder TF, Tuscano J, Sato S, Kehrl JH (1997) CD22, a B lymphocyte-specific adhesion molecule that regulates antigen receptor signaling. Ann Rev Immunol 15:481–504

Terashima M, Kim K-M, Adachi T, Nielsen PJ, Reth M, Kohler G, Lamers MC (1994) The IgM antigen receptor of B lymphocytes is associated with prohibitin and a prohibitin-related protein. EMBO J 13:3782–3792

Tisch R, Roifman CM, Hozumi N (1988) Functional differences between immunoglobulins M and D expressed on the surface of an immature B cell line. Proc Natl Acad Sci USA 85:6914–6918

Tomayko MM, Cancro MP (1998) Long-lived B cells are distinguished by elevated expression of A1. J Immunol 160:107–111

Tsui HW, Siminovitch KA, de Souza L, Tsui FW (1993) Motheaten and viable motheaten mice have mutations in the haematopoietic cell phosphatase gene. Nat Gen 4:124–129

Tuscano JM, Engel P, Tedder TF, Agarwal A, Kehrl JH (1996) Involvement of p72syk kinase, p53/p56lyn kinase and phosphatidyl inositol-3 kinase in signal transduction via the human B lymphocyte antigen CD22. Eur J Immunol 26:1246–1252

Tuveson DA, Carter RH, Soltoff SP, Fearon DT (1993) CD19 of B cells as a surrogate kinase insert region to bind phosphatidylinositol 3-kinase. Science 260:986–989

Uckun FM, Burkhardt AL, Jarvis L, Jun X, Stealey B, Dibirdik I, Myers DE, Tuel-Ahlgren L, Bolen JB (1993) Signal transduction through the CD19 receptor during discrete developmental stages of human B-cell ontogeny. J Biol Chem 268:21172–21184

van Noesel CJ, Lankester AC, van Lier RA (1993a) Dual antigen recognition by B cells. Immunol Today 14:8–11

van Noesel CJ, Lankester AC, van Schijndel GM, van Lier RA (1993b) The CR2/CD19 complex on human B cells contains the src-family kinase Lyn. Int Immunol 5:699–705

Vies DJ, Sorenson CM, Shutter JR, Korsmeyer SJ (1993) Bcl-2-deficient mice demonstrate fulminant lymphoid apoptosis, polycystic kidneys, and hypopigmented hair. Cell 75:229–240

Vitetta ES, Cambier JC, Ligler FS, Kettman JR, Uhr JW (1977) B-cell tolerance. IV. Differential role of surface IgM and IgD in determining tolerance susceptibility of murine B cells. J Exp Med 146: 1804–1808

Webb CF, Nakai C, Tucker PW (1989) Immunoglobulin receptor signaling depends on the carboxyl terminus but not the heavy-chain class. Proc Natl Acad Sci USA 86:1977–1981

Weng WK, Jarvis L, LeBien TW (1994) Signaling through CD19 activates vav/mitogen-activated protein kinase pathway and induces formation of a CD19/vav/phosphatidylinositol 3-kinase complex in human B cell precursors. J Biol Chem 269:32514–32521

Wu M, Arsura M, Bellas RE, Fitzgerald MJ, Lee H, Schauer SL, Sherr DH, Sonenshein GE (1996a) Inhibition of c-myc expression induces apoptosis of WEHI 231 murine B cells. Mol Cell Biol 16: 5015–5025

Wu M, Lee H, Bellas RE, Schauer SL, Arsura M, Katz D, Fitzgerald MJ, Rothstein TL, Sherr DH, Sonenshein GE (1996b) Inhibition of NF-κB/Rel induces apoptosis of murine B cells. EMBO J 15:4682–4690

Yang E, Zha J, Jockel J, Boise LH, Thompson CB, Korsmeyer SJ (1995) Bad, a heterodimeric partner for Bcl-XL and Bcl-2, displaces Bax and promotes cell death. Cell 80:285–291

Yao L, Kawakami Y, Kawakami T (1994) The pleckstrin homology domain of Bruton tyrosine kinase interacts with protein kinase C. Proc Natl Acad Sci USA 91:9175–9179

Yellen AJ, Glenn W, Sukhatme VP, Cao X, Monroe JG (1991) Signaling through surface IgM in tolerance- susceptible immature murine B lymphocytes. J Immunol 146:1446–1454

# Signaling Pathways that Control V(D)J Recombination

S. DESIDERIO and J. LEE



Department of Molecular Biology and Genetics, Howard Hughes Medical Institute, The Johns Hopkins University School of Medicine, Baltimore, MD 21205, USA

# 1 Introductory Remarks

The central event in the generation of immunological diversity is the assembly of antigen-receptor genes from discrete DNA segments during lymphocyte development. This process, termed V(D)J recombination, is the only known form of site-specific DNA rearrangement in vertebrates. While all immunoglobulin (Ig) and T-cell receptor (TCR) loci are assembled by means of similar reactions, recombination is highly regulated during development, both with respect to the expression of recombinase activity and the susceptibility of individual loci to rearrangement. Such control, which is critical for the enforcement of allelic exclusion and perhaps also for maintaining fidelity of rearrangement, is exerted at multiple levels, including: (1) transcription of recombinase genes, (2) accumulation of recombinase gene products, and (3) accessibility of particular loci to recombinase activity.

In this review, we will summarize what is presently known about the signaling pathways that underlie each of these levels of regulation. We will begin with an overview of lymphocyte development, after which we will describe the individual targets of control and the signaling molecules that act on them. Our goal will be to outline specific pathways that modulate V(D)J recombination and to indicate, where possible, the physiologic significance of these regulatory circuits.

# 2 Overview of V(D)J Recombination

## 2.1 Organization of Antigen Receptor Loci

The variable domains of TCR and Ig chains are encoded in discrete gene segments (V, D, and J), which are joined during lymphocyte development (Lewis 1994). V(D)J recombination exhibits exquisite specificity with respect to cell lineage and developmental stage. As T and B cells share a common recombination machinery, specificity is defined by locus-specific elements that act while in either a *cis* or *trans* conformation.

The mouse Ig-heavy-chain locus contains four $J_H$ segments, which reside about 7kb 5′ of the $C_\mu$ exons. On the order of 100 $V_H$ segments lie approximately 100kb 5′ of the $J_H$ segments and about 12 D segments are arrayed between the $V_H$ and $J_H$ clusters. At the murine Igκ-light-chain locus, four functional $J_\kappa$ segments lie about 2kb 5′ of the $C_\kappa$ exon; about 250 $V_\kappa$ segments reside to the 5′ side of the $J_\kappa$ segments. In the mouse, the Igλ locus is rather less diverse than κ, containing two Vλ segments and four Jλ-Cλ units (Blackwell and Alt 1989).

At the TCRβ locus, two nearly identical $C_\beta$ regions are tandemly arranged; a single $D_\beta$ segment and six $J_\beta$ segments lie at the 5′ side of each. About 30 $V_\beta$ segments are known; all but $V_\beta$14 are located 5′ to the $D_\beta$-$J_\beta$-$C_\beta$ array. In its organization, the TCR$\gamma$ locus resembles that of Ig$\lambda$. Three distinct, functional $J_\gamma$-$C_\gamma$ units are found in most mouse strains; only six $V_\gamma$ gene segments have been identified. The TCR$\alpha/\delta$ locus is unique in that many of the TCR$\delta$ segments are located between $\alpha$ segments and undergo deletion during $V_\alpha$ to $J_\alpha$ rearrangement. About 100 $V_\alpha$ regions have been identified, but only about 10 $V_\delta$ segments are known; the latter are split between fetal and adult usage (Davis and Bjorkman 1988).

## 2.2 The Recombination Reaction

V(D)J recombination is mediated by heptamer and nonamer signal sequences, separated by spacer regions of 12 bp or 23 bp (Lewis 1994), and is initiated by the recombination-activating proteins (RAG)-1 and RAG-2 (Schatz et al. 1989; Oettinger et al. 1990), which act in concert to cleave DNA at the junctions between antigen-receptor-coding segments and conserved recombination signals that specify sites of recombination. DNA cleavage occurs in two steps (McBlane et al. 1995). First, one DNA strand is nicked between the recombination signal sequence (RSS) heptamer and the coding sequence. This is followed by a transesterification reaction in which the free hydroxyl group at the 3′ end of the coding sequence attacks a phosphodiester on the opposite strand (van Gent et al. 1996). As a result, two DNA ends are produced: a signal end, terminating in a blunt, 5′-phosphorylated, double-strand break, and a coding end, terminating in a hairpin (Roth et al. 1992, 1993; Schlissel et al. 1993; McBlane et al. 1995; van Gent et al. 1995, 1996). Subsequent coding-to-coding and signal-to-signal end joining requires the activities of at least five proteins that also function in general double-strand DNA break repair (see Grawunder et al. 1998 for review). A sixth protein, terminal deoxynucleotidyl transferase (TdT), is responsible for the introduction of untemplated nucleotides (N regions) at coding junctions (Komori et al. 1993). N-region diversification is a developmentally regulated process that correlates with the differential expression of TdT, which is absent in fetal and newborn mice and restricted to early T and pro-B cells in adults (Li et al. 1993).

RAG-1 and RAG-2 are both essential for initiation of V(D)J recombination. Studies employing a one-hybrid assay (Difilippantonio et al. 1996) and surface plasmon resonance (Spanopoulou et al. 1996) have suggested that RAG-1 mediates RSS recognition primarily through the nonamer. RAG-2 alone has no detectable DNA-binding activity in vitro (Spanopoulou et al. 1996; Hiom and Gellert 1997), but collaborates with RAG-1 in RSS recognition (Difilippantonio et al. 1996; Hiom and Gellert 1997; Li et al. 1997). While RSS interactions with RAG-1 alone are insensitive to heptamer mutation (Difilippantonio et al. 1996; Spanopoulou et al. 1996), activities requiring both RAG-1 and RAG-2 show strong heptamer dependence (Difilippantonio et al. 1996; Hiom and Gellert 1997). Chemical interference and footprinting have shown that RAG-1 alone is able to make extensive contacts with the nonamer, but occupancy of the heptamer requires the participation of RAG-2 (Swanson and Desiderio 1998). The requirement for collaboration between RAG-1 and RAG-2 in RSS recognition implies that V(D)J recombination can be controlled at its earliest stages by regulating the expression or activity of either protein.

# 3 Levels of Control

## 3.1 Developmental Stage- and Lineage-Specific Regulation of V(D)J Recombination at Individual Loci

### 3.1.1 Lineage Specificity

Assembly of Ig and TCR genes is tissue-specifically regulated; thus, neither Ig nor TCR genes are assembled in non-lymphoid cells. Furthermore, antigen-receptor gene assembly exhibits lineage specificity, in that Ig genes are assembled completely only in B lineage cells, while TCR genes are only assembled completely in T lymphocytes. T cells can be subdivided into two distinct lineages based on the expression of αβ or γδ forms of the TCR. The successful completion of either TCRγδ- or β gene rearrangements may decide the lineage commitment into αβ or γδ T cell.

### 3.1.2 Developmental Regulation and Allelic Exclusion

With respect to V(D)J recombination, the pathways of B-cell and αβ T-cell development share several features (Fig. 1). First, recombination occurs in two discrete waves: rearrangement of antigen-receptor heavy-chain (TCRβ or Igμ) genes generally precedes that of the light-chain (TCRα or IgL) genes. Second, functional rearrangement of a heavy-chain gene is followed by construction of an immature form of the antigen receptor, in which TCRβ or Igμ is associated with

**Fig. 1a,b.** Coordination of lymphocyte development with recombination-activating gene (*RAG*) expression. **a** Comparison of αβ T-cell and B-cell development. In both the αβ T and B lineages, productive rearrangement of an antigen receptor heavy-chain gene [T cell receptor (TCR)β or immunoglobulin (Ig)μ] is succeeded by appearance of a developmental subset ($CD4^-CD8^-CD25^-CD44^{lo}$ and $CD45R^+CD25^+CD43^+$, respectively), which undergoes rapid cycling and proliferative expansion. Subsequently, these cells give rise to a resting population in which most light-chain rearrangement (TCRα or IgL) occurs. **b** During the stages of B-cell development prior to the appearance of sIg, *RAG* transcripts are expressed in two waves. *RAG* transcripts are downregulated upon the proB/preB-I to large preB-II transition. *RAG* transcripts reappear during the small preB-II stage and are downregulated again after successful light-chain rearrangement. Ligation of sIg on immature B ($IgM^+IgD^-$) cells opposes downregulation of *RAG* transcripts at this stage, thereby permitting receptor editing. A third wave of *RAG* transcription has been reported in so-called neotenic germinal center B cells of immunized mice. β/Ψα and μ/Ψ*L* indicate a complex of TCRβ or Igμ with the surrogate α or L chain, respectively. For references, see text

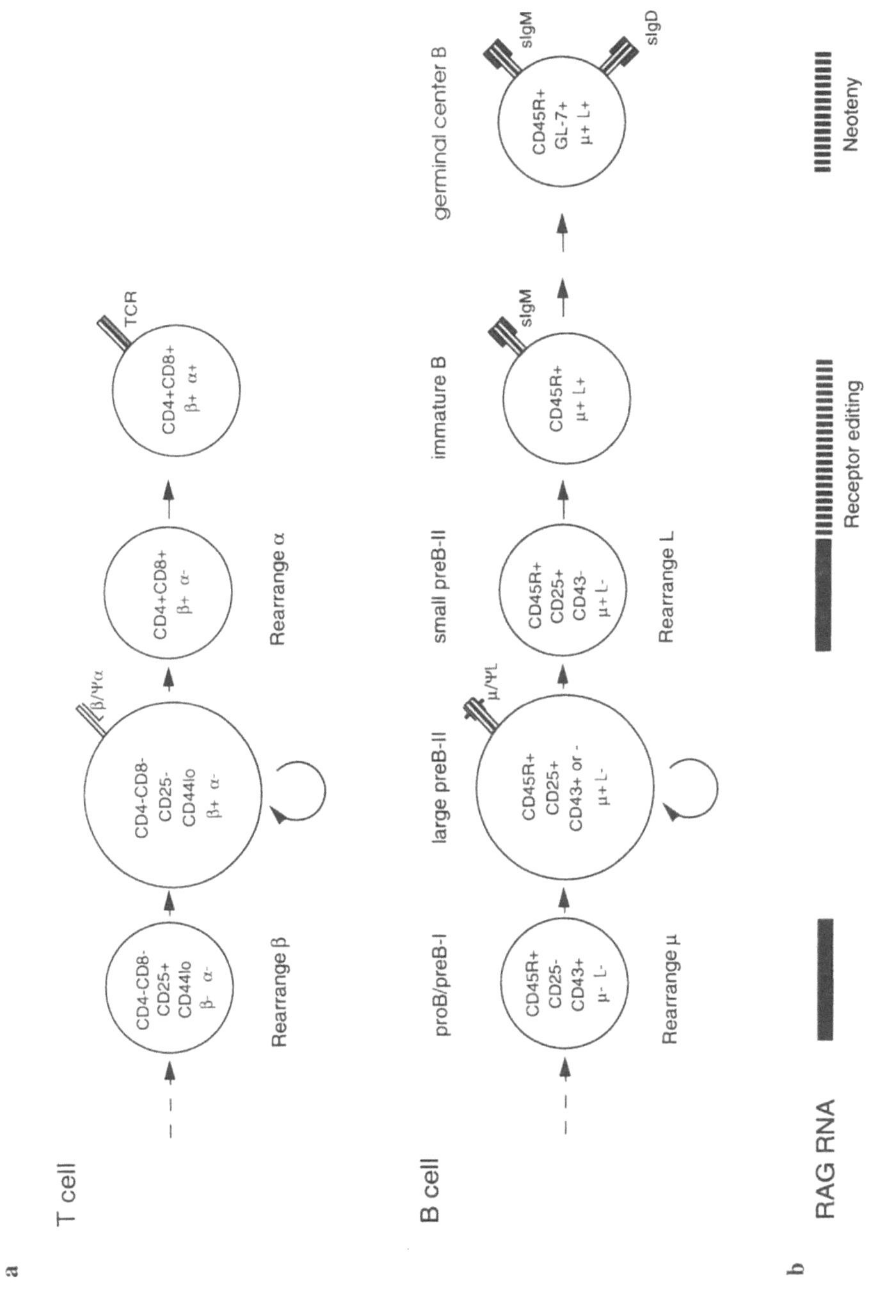
a
T cell
CD4-CD8- CD25+ CD44lo β- α-
Rearrange β
CD4-CD8- CD25- CD44lo β+ α-
β/Ψα
CD4+CD8+ β+ α-
Rearrange α
CD4+CD8+ β+ α+
TCR
B cell
proB/preB-I
CD45R+ CD25- CD43+ μ- L-
Rearrange μ
large preB-II
CD45R+ CD25+ CD43+ or - μ+ L-
μ/ΨL
small preB-II
CD45R+ CD25+ CD43- μ+ L-
Rearrange L
immature B
CD45R+ μ+ L+
sIgM
germinal center B
CD45R+ GL-7+ μ+ L+
sIgM
sIgD
b
RAG RNA
Receptor editing
Neoteny

invariant, surrogate light chains (TCRβ with pre-Tα; Igμ with VpreB and λ5). Third, the appearance of immature antigen-receptor forms is accompanied by a period of rapid cellular expansion, during which further V(D)J rearrangements at the heavy-chain loci are suppressed (allelic exclusion). Fourth, this proliferative phase is followed by a quiescent period, during which light-chain gene rearrangement (TCRα or IgL) occurs (see LIN and DESIDERIO 1995 for review).

These analogies are evident when we consider the relationship of V(D)J rearrangement to cellular maturation and expansion during lymphocyte development (BOYER et al. 1989; DECKER et al. 1991; ALT et al. 1992; GODFREY et al. 1993; GROETTRUP et al. 1993; GROETTRUP and VON BOEHMER 1993; ROLINK et al. 1993; SHINKAI et al. 1993; DUDLEY et al. 1994; GODFREY et al. 1994; KARASUYAMA et al. 1994; ROLINK et al. 1994; SAINTRUF et al. 1994; SPANOPOULOU et al. 1994; WILSON et al. 1994). In the B lineage, rearrangement of the Igμ locus occurs in the slowly dividing pro-B/pre-B-I population, which expresses the phenotype $CD45R^+CD25^-CD43^+$. Productive Igμ rearrangement coincides with the transition to the $CD45R^+CD25^+$ (large pre-B-II) stage. During this developmental transition, which requires expression of a functional Igμ chain, B-cell progenitors are induced to undergo rapid expansion, V(D)J recombination is suppressed, and expression of CD43 is lost. These large, dividing, pre-B-II cells subsequently enter a resting phase (small pre-B-II), during which Ig L-chain gene rearrangement occurs. Productive light-chain rearrangement is accompanied by the appearance of surface Ig (sIg) and cessation of V(D)J recombination.

In the T lineage, TCRβ gene rearrangement occurs in the $CD4^-CD8^-$ (double negative) thymocyte compartment. This population, which can be subfractionated with respect to expression of CD44 and the interleukin 2 (IL-2) receptor (CD25), matures in the following sequence: $CD44^{hi}CD25^- \rightarrow CD44^{hi}CD25^+ \rightarrow CD44^{lo}CD25^+ \rightarrow CD44^{lo}CD25^-$ (GODFREY et al. 1993); TCRβ rearrangement is activated following the progression from $CD44^{hi}CD25^+$ to $CD44^{lo}CD25^+$. The $CD44^{lo}CD25^+ \rightarrow CD44^{lo}CD25^-$ transition, which is accompanied by cessation of further rearrangement at the TCRβ locus and a ten-fold increase in total thymocyte number, requires expression of a functional TCRβ chain. This expansion is followed by the emergence of a nondividing $CD4^+CD8^+$ (double positive) population and commencement of TCRα rearrangement. Downregulation of V(D)J recombination follows construction of a functional TCRα gene and expression of an αβ-TCR complex at the cell surface.

Thus, V(D)J rearrangement is regulated as a function of multiple but interdependent physiologic variables, including cell lineage, developmental stage and antigen-receptor expression. The linkage of recombination to changes in these physiologic states has important functional consequences, including the assembly of B- or T-cell-specific receptor types, enforcement of allelic exclusion, coordination of RAG activity with double-strand-break repair, and the editing of antigen-receptor specificities.

## 3.2 Control of *RAG* RNA Accumulation

### 3.2.1 Expression of *RAG* Transcripts during Lymphocyte Development

The *RAG*-1 and *RAG*-2 genes are closely linked in all species in which they have been identified, probably reflecting their evolutionary origin as components of a transposable DNA element (AGRAWAL et al. 1998). Their distance ranges from 6kb in xenopus to 8kb in the mouse; in the latter, the *RAG* genes are located on chromosome 2. *RAG*-1 and *RAG*-2 are transcribed convergently. They are coexpressed at the RNA level except in avian bursa, where *RAG*-2 is transcribed in the absence of *RAG*-1 (CARLSON et al. 1991), and possibly in the central nervous system, in which one group has described low-level accumulation of *RAG*-1 transcripts in the absence of *RAG*-2 (CHUN et al. 1991).

The *RAG* genes are expressed in two discrete waves during development of the T and B lineages in the primary lymphoid tissues. *RAG*-1 and *RAG*-2 transcripts accumulate preferentially in the DN $CD44^{lo}CD25^{+}$ and DP thymocyte populations, in which TCR β and α rearrangements occur (WILSON et al. 1994). Recombinase activity is extinguished upon subsequent engagement of the αβ TCR. Similarly, in the B lineage, *RAG* transcripts are preferentially found in proB and preB-I cells, which undergo D to $J_H$ and $V_H$ to $DJ_H$ rearrangement and in small preB-II cells, in which IgL-chain rearrangement occurs (KARASUYAMA et al. 1994; ROLINK et al. 1994). In contrast to events at the comparable stage of T-cell development, engagement of sIg on immature B cells, typically by self-antigen, results in persistence of *RAG* transcription and continuing Ig gene rearrangement, which permits the editing of receptor specificities (GAY et al. 1993; TIEGS et al. 1993).

A third wave of *RAG* expression has been reported to occur in germinal-center (GC) B cells of the mouse (HAN et al. 1996; HIKIDA et al. 1996). These cells share several phenotypic characteristics with immature B cells but can be distinguished by their expression of peanut agglutinin (PNA) and their reactivity to the monoclonal antibody GL-7. RAG-1 and RAG-2 are detectable in $PNA^{+}$, $GL\text{-}7^{+}$ splenic GC B cells after immunization and in Peyer's-patch GC B cells of nonimmunized mice. The antigen-driven activation of *RAG* expression in GC B cells is associated with the appearance of $J_H$-associated DNA breaks and the accumulation of otherwise allelically excluded gene rearrangements in Ig transgenic mice (PAPAVASILIOU et al. 1997). These observations have been interpreted to represent the re-expression of RAG proteins in mature B cells and a partial return to an immature state, termed neoteny (HAN et al. 1996), in which receptor specificities may be altered. Alternative explanations exist, however, including the possibility that *RAG*-expressing GC cells may represent an immature cell type circulating through the GC. Thus, the physiologic significance of *RAG* expression in GC cells is as yet unresolved. Evidence has recently emerged suggesting that a third wave of *RAG* transcription, accompanied by TCR gene rearrangement, also occurs in peripheral $CD4^{+}$ T cells (MCMAHAN and FINK 1998).

### 3.2.2 Putative *Cis*- and *Trans*-Acting Regulators of *RAG* Transcription

An understanding of *RAG* gene expression and its control would require identification of DNA sequence elements that govern *RAG* transcription and the proteins that interact with them. Unfortunately, progress toward these goals has been slow. In human and mouse B lymphoid cells, *RAG*-1 is transcribed from multiple start sites, clustered within a region of about 30 bp (KURIOKA et al. 1996; FULLER and STORB 1997). Neither a consensus TATA box nor an initiator sequence is present in this interval. A DNA fragment from the human *RAG*-1 upstream region has been reported to support expression of a transfected reporter construct; activity was dependent on the interval from −123 bp to −86 bp (KURIOKA et al. 1996), which contains a functional binding site for the transcription factor nuclear factor (NF)-Y. This putative promoter does not exhibit cell-type specificity and is active in *RAG*-1-expressing and non-expressing cell lines. A similar fragment from the murine *RAG*-1 gene, extending from −205 bp to +75 bp, did not support expression in pre-B, pre-T or mature lymphoid cell lines unless the IgH-chain enhancer was included (FULLER and STORB 1997). Under these conditions, mutation of the NF-Y site profoundly impaired promoter function and NF-Y binding in vitro; on this basis it has been proposed that NF-Y may recruit additional transcription factors to the *RAG*-1 promoter, by analogy to its function in the transcription of major histocompatibility complex (MHC) class-II genes.

In the absence of evidence for developmental-stage- or tissue specificity, several attempts have been made to correlate putative regulatory sequences with DNase-I-hypersensitive sites in vivo (KITAGAWA et al. 1996; FULLER and STORB 1997). In the vicinity of the human *RAG*-1 gene, three DNase-I-hypersensitive sites have been detected in *RAG*-1-expressing cell lines but not in non-expressing cells (KITAGAWA et al. 1996). Two of these are upstream of the transcription start sites; the third is within the *RAG*-1 intron. Removal of the first site, at about −0.8kb relative to the 5′ ends of the *RAG*-1 transcripts, had no effect on promoter activity in Nalm-6 cells; the second site lies near the NF-Y recognition site, but whether DNase hypersensitivity is dependent on binding of NF-Y remains unclear. None of these sites is associated with enhancer activity. In mouse cells expressing *RAG*-1, a strong DNase-I-hypersensitive site has also been located near the transcriptional start points, but, as in the case of the human gene, the relationship of this site to the binding of specific transcription factors has not yet been established (FULLER and STORB 1997).

Gene-ablation studies have identified several transcription factors that are essential for normal lymphoid development. Disruption of the gene encoding early B-cell factor (LIN and GROSSCHEDL 1995) or E2A (BAIN et al. 1994; ZHUANG et al. 1994) blocks B-cell development at the proB/preB-I stage, prior to the initiation of V(D)J rearrangement at the heavy-chain locus. Although the absence of *RAG* expression in E2A-deficient B-cell progenitors has suggested a role for E2A in the regulation of *RAG* transcription (BAIN et al. 1994), the data are equally consistent with blockage of B-cell development prior to activation of the *RAG* genes. Homozygous disruption of the gene for the transcription factor Pax5/BSAP also impairs B-cell development, but in this instance the block occurs after the initial expression of *RAG*-1 and *RAG*-2, indicating that Pax5/BSAP is dispensable for *RAG* transcription in early B-cell development and in the T lineage, whose development is unaffected (URBANEK et al. 1994). Notably, Pax-5 deficiency impairs V to DJ but not D to J rearrangement, suggesting a role in the control of locus accessibility to the recombinase. Available data permit the possibility that Pax5/BSAP may play a role in the regulation of *RAG* transcription at later times in B-cell development.

### 3.2.3 Pharmacologic Modulation of *RAG* Transcripts

The steady-state levels of *RAG*-1 mRNA and the efficiency of V(D)J recombination, as assayed by rearrangement of an extrachromosomal substrate, can be coinduced by pharmacologic elevators of cyclic adenosine monophosphate (cAMP), such as caffeine and theophylline; conversely, accumulation of *RAG*-1 transcripts and the frequency of recombination are suppressed by antagonism of cAMP by treatment with the phorbol ester phorbol myristate acetate and the calcium ionophore A23187 (MENETSKI and GELLERT 1990). The effect of phorbol ester appears to be exerted at the transcriptional and post-transcriptional levels, as treatment of pre-T or pre-B cell lines with the phorbol ester phorbol 12-tetradecanoate 13-acetate is associated with rapid elimination of steady-state *RAG*-1 and *RAG*-2 mRNA, cessation of *RAG*-1 transcription, and rapid degradation of existing *RAG*-1 transcripts (NEALE et al. 1992).

## 3.3 Post-Translational Regulation of RAG Activity

### 3.3.1 Periodic Degradation of RAG-2 Protein in Cycling Cells

RAG-1 and RAG-2 are both essential for initiation of V(D)J recombination, and this process does not proceed in the absence of either protein. The level of RAG-2 protein in developing lymphocytes exhibits a striking cell-cycle dependence, accumulating preferentially in G0/G1 cells and disappearing to low or undetectable levels during the S and G2/M cell-cycle phases (LIN and DESIDERIO 1994). The oscillation of RAG-2 protein in dividing lymphoid cells is unaccompanied by corresponding changes in the steady-state levels of *RAG*-2 transcripts (LIN and DESIDERIO 1994), indicating that periodicity of RAG-2 expression is enforced at the post-transcriptional level.

RAG-2 accumulation is governed by signals residing within the carboxy-terminal 89 amino acid residues of the protein (LIN and DESIDERIO 1993). The major determinants of RAG-2 instability consist of a minimal cyclin-dependent kinase (CDK) phosphorylation site at residues 490 and 491 and a cationic interval extending from residues 499 through 508. A deletion that removes the cationic region but spares the CDK phosphorylation site is associated with accumulation of a phospho-Thr490 form of RAG-2 in vivo, consistent with the idea that phosphorylation of RAG-2 by one or more CDKs targets it for rapid degradation (LIN and DESIDERIO 1993; LI et al. 1996). Accumulation of *RAG*-2 in the cell cycle is governed by periodic degradation; mutations that prolong the half-life of RAG-2 protein abolish its disappearance at the G1-S boundary (Li et al. 1996).

### 3.3.2 RAG-2 Accumulation and the Timing of V(D)J Recombination

Accumulation of V(D)J recombination intermediates shows a pattern of cell-cycle dependence similar to that of RAG-2. In thymocytes, Ig $J_H$ and TCR Dδ2 signal ends are recovered preferentially from quiescent and G1-phase cells (Schlissel et al. 1993; Desiderio et al. 1996). Through the use of transgenic mice that express wild-type or Thr490 mutant RAG-2 protein in the thymus, the cell-cycle-dependent expression of RAG-2 protein has been shown to drive the periodic accumulation of V(D)J-recombination intermediates (Li et al. 1996). In RAG-2-deficient mice reconstituted with wild-type transgenic RAG-2, signal-end intermediates accumulate exclusively in G0/G1-phase thymocytes. Reconstitution with the RAG-2 T490 mutant, in contrast, permits accumulation of signal ends throughout the cell cycle. These observations support a model (Lin and Desiderio 1995) in which phosphorylation of RAG-2 by one or more CDKs links recombinase activity and initiation of V(D)J rearrangement to the cell cycle.

While this linkage may serve to prevent potential deleterious effects of RSS cleavage during DNA synthesis or mitosis, transgenic mice expressing RAG-2 constitutively in the cell cycle exhibit no increase in the occurrence of malignancy (Li et al. 1996) or interlocus rearrangements (Lee and Desiderio, unpublished observations). Another possible role for the cell-cycle regulation of RAG activity is suggested by a consideration of the mechanism by which V(D)J rearrangement is completed. There are two general mechanisms for the repair of double-stranded-DNA breaks in mammalian cells: non-homologous end joining (NHEJ) and homologous recombination; the NHEJ pathway, which is essential for completion of V(D)J recombination, appears to dominate during the G1 and early S phases (Maser et al. 1997; Takata et al. 1998). The temporal coincidence of RAG-2 accumulation and NHEJ activity may serve to coordinate RSS-associated DNA cleavage with the completion of V(D)J recombination by NHEJ.

# 4 Specific Pathways for Control of Antigen-Receptor Assembly

## 4.1 Control of Recombinase Activity

### 4.1.1 Control by Lymphokines and Stromal Factors

#### 4.1.1.1 Interleukin-7 (IL-7)

Stromal cells mediate development of B-cell progenitors through direct contact and secreted factors such as IL-7, an essential growth factor for B-lineage precursors in mice (see Candéias et al. 1997 for review). While the development of early pro-B cells is independent of IL-7, late pro-B and early pre-B cells express IL-7R, and IL-7 is required for the proliferative expansion of early pre-B-II ($\mu^+ \psi L^+$) cells. In contrast, late pre-B-II, immature B and mature B cells lack functional IL-7R and are independent of IL-7.

In B-cell progenitors, an inverse relationship exists between the IL-7-mediated signaling and V(D)J recombination (Rolink et al. 1991a,b; Billips et al. 1995; ten Boekel et al. 1995). Early B-cell clones that were maintained on stromal cells in the presence of IL-7 were found to carry $DJ_H$ rearrangements and germline light-chain loci; removal of IL-7 was accompanied by cessation of proliferation and differentiation to $sIg^+$ B cells (Rolink et al. 1991a,b). Treatment of human pro-B cells with IL-7 suppresses expression of *RAG*-1, *RAG*-2 and TdT (Billips et al. 1995) and induces expression of CD19 (Wolf et al. 1993; Billips et al. 1995), a transmembrane protein that regulates B-lymphoid proliferation; TdT downregulation precedes CD19 induction, which is followed by downregulation of *RAG* expression (Billips et al. 1995). Notably, the negative effect of IL-7 on *RAG* expression could be reversed by ligation of CD19 (Billips et al. 1995). Based on these observations, a model has been proposed in which the *RAG* genes would continue to be expressed in the presence of IL-7 as long as CD19 ligation persists; as cells are displaced from the stroma as a consequence of proliferative expansion, CD19 would no longer be ligated and *RAG* expression would decrease (Billips et al. 1995). These results suggest, moreover, that IL-7 is able to modulate the immune repertoire by way of its action on TdT.

X-linked agammaglobulinemia (XLA) is characterized by a block in B-cell development at the pre-B cell stage and is caused by a deficiency in the tyrosine kinase Btk (Tsukada et al. 1993; Vetrie et al. 1993). In patients with XLA, TdT expression persists abnormally in pre-B cells, suggesting a defect in its negative regulation. Accordingly, IL-7 did not downregulate TdT expression in cultured pro-B cells from XLA patients; moreover, ligation of CD19 was unable to abrogate the suppression of *RAG* expression by IL-7. These observations have suggested that the tyrosine kinase Btk, which is defective in subjects with XLA, participates in the reciprocal modulation of *RAG* activity by IL-7 and CD19 (Billips et al. 1995).

In contrast to its suppressive effect on V(D)J recombination in immature B cells, IL-7, with anti-CD40, has been reported to induce re-expression of *RAG*-2 and the appearance of Ig$\lambda$ excision products in mouse $IgD^+$ B cells in vitro (Hikida and Ohmori 1998); similar observations have been made for IL-4 (see below). In addition, antibody-mediated blockade of the IL-7R suppressed expression of *RAG*-2 and V(D)J recombination in draining lymph nodes of immunized mice (Hikida et al. 1998). While this result is somewhat puzzling, as mature B cells lack IL-7R, reappearance of the IL-7R has been reported when such cells are stimulated in vitro with IL-7 and anti-CD40. On the basis of these observations, it has been proposed that IL-7 plays a physiologic role in the induction of V(D)J recombination in GC B cells (Hikida et al. 1998), although the distinction between the effect of IL-7 on mature cells and the opposite effects seen on immature cells (above) should be noted. One should cite in this regard a report that IL-7 induces TCR$\beta$ gene rearrangement and sustained expression of *RAG* genes in mouse fetal thymic-organ cultures (Muegge et al. 1993), although the complexity of the organ-culture system admits the possibility of indirect effects.

#### 4.1.1.2 Interleukin-4 (IL-4)

Several recent studies provide evidence that IL-4, in combination with certain other stimuli, can reproduce features of neoteny in mature B cells. Activated, mature $sIgD^+$ B cells can be induced to re-express *RAG*-1 and *RAG*-2 in vitro by IL-4 in combination with lipopolysaccharide (LPS) or anti-CD40 antibody (Hikida et al. 1996). Treatment with IL-4 and LPS results in the appearance of $J_H$-associated double-stranded-DNA breaks in mouse splenic B cells and the deletional replacement of prerearranged Ig genes (Papavasiliou et al. 1997). Furthermore, the induction of *RAG* expression on treatment with IL-4 and LPS is accompanied by the appearance of Vλ-Jλ circular excision products, indicative of ongoing V(D)J recombination (Hikida and Ohmori 1998). Thus, IL-4, in combination with LPS or anti-CD40, appears to mimic the reinduction of *RAG* expression observed in GC B cells or draining lymph nodes of immunized mice. Nonetheless, if IL-4 plays a physiologic role in this process, it must be a redundant one, because IL-4 deficiency does not impair *RAG* expression in draining lymph nodes of immunized mice. One possible source of redundancy is IL-7, which, in combination with anti-CD40, can also induce *RAG* expression in mature B cells (see above).

#### 4.1.1.3 Stromal Culture Systems

IL-3, IL-6 and IL-7 have been reported to induce *RAG*-1 transcripts synergistically in human bone-marrow cells cocultured with the mouse stromal cell line PA-6 (Tagoh et al. 1996). Because four- to fivefold increases were attained after 24 h of treatment, it is difficult to rule out the possible outgrowth of *RAG*-expressing cells in these experiments. More convincing is a similar result obtained with the FL8.2.4.4 human lymphoid progenitor cell line, in which the observed increase of 40- to 50-fold in steady-state *RAG* RNA was incompatible with the doubling time. Contact between FL8.2.4.4 cells and the PA-6 stromal layer was essential for induction of *RAG*-1 RNA; fixation of the PA-6 monolayer did not impair its ability to support induction. While this study underscores the possibility that soluble and insoluble factors may collaborate in the control of *RAG*-1 expression, the physiologic significance of these observations is difficult to assess, not least because the recombination frequency in electroporated induced FL8.2.4.4 cells was very low relative to control.

One recent study has attempted to model the induction of *RAG* expression in primitive human hematopoietic progenitors (Gaffney et al. 1998). In such cells, purified on the basis of their $CD34^+Lin^-DR^-$ phenotype, transcripts for CD3ε, Ikaros, CD10, TdT, and *RAG*-1 were observed. Transcripts for *RAG*-2, CD3γ, CD3δ, and CD3ζ were absent. *RAG*-2 transcripts were detectable after culture for 14 days in the presence of IL-2, IL-3, IL-7, c-kit ligand, and stroma-conditioned medium (SCM). *RAG*-2 RNA was detectable at an earlier time (i.e., 7 days) when FLT-3 ligand was included with the other culture components. As with the aforementioned study, however, it is difficult, given these observations, to distinguish induction of *RAG*-2 expression (as a marker for differentiation) from the outgrowth of a subpopulation.

### 4.1.2 Modulation of *RAG* Transcription by Antigen-Receptor Engagement

The relationship between engagement of antigen-receptor complexes and expression of the *RAG* genes is complex. Depending on the developmental stage of the lymphocyte, antigen-receptor stimulation can result in upregulation or downregulation of *RAG* transcripts and concomitant induction or suppression of recombination. These distinct relationships between receptor engagement and *RAG* transcription are particularly well illustrated in the B lineage. In immature B lymphocytes, B-cell-receptor (BCR) signaling stimulates the editing of receptor specificities by continued V(D)J recombination. Receptor editing, which provides a means to achieve immune tolerance, can occur by rearrangement of upstream V segments to downstream J segments at light-chain loci, or by frank replacement of one V segment by another. Transgenic-mouse models indicate that this process is initiated by an interaction between surface Ig and self antigen, which is accompanied by accumulation of *RAG* transcripts and V(D)J recombination intermediates in immature ($IgM^{+}IgD^{-}$) bone-marrow B cells (Hertz and Nemazee 1997). The ability of transgenic Ig to circumvent the inhibition of Ig gene rearrangement and maturation of B-cell precursors by IL-7 (Melamed et al. 1997) is consistent with stimulation of V(D)J recombination by the BCR in immature B cells.

A converse relationship between BCR engagement and V(D)J recombination is seen in that subpopulation of splenic B cells in which immunization or treatment in vitro with IL-4 and LPS induces re-expression of *RAG* transcripts and V(D)J recombination. In these cells, BCR ligation inhibits the induction of V(D)J recombination by IL-4 and LPS (Hertz et al. 1998). Furthermore, in mice expressing an antigen-specific transgenic Ig, immunization with a low-affinity ligand is able to induce V(D)J recombination in the periphery, while high-avidity ligands do not, suggesting that the reappearance of V(D)J recombination in mature B cells selectively modifies antigen receptors with weak binding (Hertz et al. 1998).

### 4.1.3 Cyclin/Cdk-Mediated Control of Recombinase Activity

As outlined earlier, V(D)J recombination is coupled to the cell cycle by periodic destruction of RAG-2 protein. The amino-acid residue whose phosphorylation triggers degradation, Thr490, lies within a CDK consensus substrate sequence, and several CDK complexes can phosphorylate this site in vitro, including cyclin E/CDK2 and cyclin A/CDK2 (Lin and Desiderio 1993). In vivo, however, accumulation of RAG-2 is specifically opposed by cyclin A/CDK2; conversely, *RAG*-2 is induced by $p27^{Kip1}$ and other CDK inhibitors (CKIs) that suppress CDK2 activity (Lee and Desiderio, unpublished observations). These results are in agreement with the observed inverse correlation between cyclin A expression and RAG-2 accumulation in immature thymocytes (Hoffman et al. 1996).

Coinduction of RAG-2 and G1 delay by $p27^{Kip1}$ was found to stimulate V(D)J recombination, but induction of RAG-2 protein outside of G1 did not increase recombination frequency (Lee and Desiderio, unpublished observations). These observations suggest pathways by which cytokines and other extracellular stimuli may regulate recombinase activity distinct from, and superimposed upon, their effects on *RAG* transcription. IL-7, for example, induces CDK2 activity and stimulates G1/S progression (Itoh et al. 1996), consistent with its ability to inhibit V(D)J recombination in B-cell progenitors. Conversely, the cytostatic factor TGF-β, which retards cell-cycle progression of $CD4^{-}CD8^{lo}$ precursor cells (Takahama et al. 1994), increases intracellular levels of free $p27^{Kip1}$ (Reynisdottir et al. 1995); this action is consistent with the assembly of heavy- and light-chain genes during periods of relatively slow cell division or cytostasis. These inferences are also consistent with earlier observations demonstrating differentiation of Bcl-2-expressing pre-B cells to $sIgM^{+}$ cells in the absence of IL-7 and in the presence of hydroxyurea, an inhibitor of DNA replication (Rolink et al. 1993).

## 4.2 Control of Locus Accessibility

In a given cell expressing V(D)J recombinase activity, differential accessibility of antigen-receptor gene segments has been hypothesized to govern the susceptibility of particular gene segments to rearrangement. Accessibility has variously been correlated with transcription, hypomethylation, and open chromatin structure at unrearranged gene segments.

### 4.2.1 Transcription

Transcription through germline antigen-receptor segments is detectable at times when these elements are undergoing recombination (see Sleckman et al. 1996 for review). Although transcriptional activity is not strictly correlated with rearrangement, a close association between these phenomena has been observed. Mutations within *cis*-acting transcriptional regulatory sites have been tested for their effects on V(D)J recombination.

#### 4.2.1.1 The Igμ Intronic Enhancer (Eμ)

The mouse Ig-heavy-chain intronic enhancer (Eμ) is located between the $J_H$ cluster and the Cμ exons. Enhancer activity has been localized to an interval containing a 95-bp core element, a germline promoter (Iμ), and flanking matrix-attachment regions (MARs). Eμ is sufficient to induce transcription and rearrangement of a transgenic mini-locus (Ferrier et al. 1990). The ability of Eμ to support recombination and transcription can be separated; in a transgenic recombination substrate, point mutations within the E-box motifs μE3 and μE4 abolished rearrangement without affecting transcription of the participating J segment (Fernex et al. 1995). These observations are consistent with the ability of the

E-box-binding protein E47 to induce $D_H$ to $J_H$ recombination when transfected into pre-T cells (SCHLISSEL et al. 1991). Targeted deletion of Eμ, however, resulted in only a partial block in V(D)J recombination (CHEN et al. 1993; SERWE and SABLITZKY 1993); the inhibitory effect on $V_H$ to $D_HJ_H$ joining was substantially more severe than that on $D_H$ to $J_H$ joining (SERWE and SABLITZKY 1993). Thus, other *cis*-regulatory elements are likely to act together with Eμ in the activation of V(D)J recombination at the Ig-heavy-chain locus.

### 4.2.1.2 The Igκ Intronic and 3′ Enhancers

The Ig-light-chain κ locus is associated with an intronic enhancer (iEκ), residing between the Jκ region and the Cκ exon and a 3′ enhancer (3′Eκ), which lies about 9kb 3′ of Cκ. Targeted cre-loxP deletion of the iEκ resulted in inhibition of $V_\kappa J_\kappa$ rearrangement and a reduction in the number of $Ig\kappa^+$ B cells (XU et al. 1996). Similarly, homozygous cre-loxP deletion of the 3′Eκ was associated with a threefold decrease in the number of splenic $Ig\kappa^+$ B cells and increased retention of the germline Jκ in $Ig\lambda^+$ B cells (GORMAN et al. 1996). While the *trans*-acting factors that regulate recombination through the Igκ enhancers are not well characterized, several studies have implicated PU.1 and NFκB in this process (DEMENGEOT et al. 1995; HIRAMATSU et al. 1995; HAYASHI et al. 1997). In transgenic substrates, mutation of a PU.1-binding motif in the 3′Eκ impaired lineage and developmental-stage specificity of rearrangement (HIRAMATSU, AKAGI et al. 1995; HAYASHI et al. 1997). In gene-ablation studies, however, deletion of the endogenous 3′Eκ did not deregulate kappa-gene rearrangement (VAN DER STOEP et al. 1998), forcing revision of the view obtained through transgenic experiments. In a recombinase-inducible cell line, transcription, demethylation, and rearrangement of an exogenous, integrated substrate was dependent on an intact κB motif in iEκ (DEMENGEOT et al. 1995). In pre-B cells, blockade of NFκB nuclear translocation inhibited activation of germline transcription, DNase-I hypersensitivity, and $V_\kappa$ to $J_\kappa$ recombination by LPS but not by interferon (IFN)-γ, thereby suggesting the existence of multiple independent pathways by which recombination is activated at the Igκ locus (O'BRIEN et al. 1997).

### 4.2.1.3 The TCRβ Enhancer

The TCRβ enhancer (Eβ) is located within a 0.5-kb DNA segment that lies 7kb 3′ of the $C_\beta 2$ region. The minimal Eβ contains binding sites for members of the cAMP response element binding/activating transcription factor (CREB/ATF), core-binding factor/polyomavirus-enhancer-binding protein 2 (CBF/PEBP2), T-cell factor-1 (TCF-1)/lymphoid enhancer-binding factor (LEF-1), and Ets families of transcription factors (CLEVERS and FERRIER 1998). The minimal Eβ is sufficient to support T-lineage-specific and temporal-stage-specific D to J recombination in a transgenic construct (CAPONE et al. 1993). A cre-loxP-mediated deletion of the 700-bp XbaI/EcoRI fragment spanning Eβ was found to decrease germline transcription by more than 1000-fold and to abolish $D_\beta$ to $J_\beta$ recombination on the mutated allele (BORIES et al. 1996). In homozygous mutant mice bearing a somewhat smaller deletion of 560 bp spanning the minimal Eβ, a 10- to 20-fold decrease in the number of total thymocytes, a severe paucity in the number of $TCR\alpha\beta^+$ thymocytes, and a greater than 100-fold decrease in $D_\beta2$-$J_\beta2$ and $V_\beta12$-$J_\beta2$ recombination was observed (BOUVIER et al. 1996). These earlier observations indicated a functional role for Eβ sequences in V(D)J recombination at the TCRβ locus.

More recent experiments have begun to define the specific aspects of V(D)J recombination that are impaired by removal of Eβ (HEMPEL et al. 1998). Contrary to expectation, the 560-bp Eβ deletion did not substantially impair the accessibility of $D_\beta$ and $J_\beta$ segments to cleavage by the recombinase, nor was formation of signal joints impaired in these animals. In contrast, $D_\beta$-$J_\beta$-coding joint formation was severely impaired, suggesting that Eβ functions to direct DNA repair at specific TCRβ sites during the late phase of the V(D)J recombination reaction. These observations force a re-evaluation of the interpretation that locus-specificity of V(D)J recombination is governed solely by accessibility.

#### 4.2.1.4 The TCRα Enhancer

The TCRα enhancer (Eα) is located 4.5kb 3′ of Cα in the human TCRα locus and is sufficient to confer tissue-, lineage-, and developmental-stage-specific VD to J recombination on a transgenic minilocus (LAUZURICA and KRANGEL 1994). Moreover, homozygous, cre-loxP-mediated deletion of Eα blocked T-cell development at the double-positive stage, and the targeted alleles of splenic αβ T cells in heterozygous mice were unrearranged in heterozygous mice (SLECKMAN et al. 1997). Like Eβ, Eα contains binding sites for TCF-1/LEF-1 and Ets (CLEVERS and FERRIER 1998). Mutation of the binding site for either of these transcription factors blocked enhancer-dependent recombination of a transgenic minilocus (ROBERTS et al. 1997), suggesting involvement of these proteins in the control of TCRα rearrangement. The potential involvement of Ets proteins in V(D)J recombination is particularly interesting because these factors are activated by microtubule-associated protein kinases and may, therefore, link external signaling pathways to antigen-receptor rearrangement. Germline deletion of the Ets-1 gene, however, had no effect on the expression of TCRαβ (BORIES et al. 1995; MUTHUSAMY et al. 1995), indicating that this particular Ets family member is dispensable for αβ T-cell development. Germline deletion of the gene for TCF-1/LEF-1, in contrast, blocks the transition from the $CD8^+$ intermediate single-positive to the $CD4^+CD8^+$ double-positive stage, suggesting that this protein is dispensable for TCRβ rearrangement but complicating direct evaluation of its role in TCRα assembly (VERBEEK et al. 1995).

#### 4.2.1.5 The TCRδ Enhancer

The TCRδ enhancer (Eδ) is located within the Jδ3–Cδ intron. Eδ is essential for VD to J joining in a TCRδ minilocus transgene (LAUZURICA and KRANGEL 1994). The minimal Eδ core, δE3, contains adjacent binding sites for the transcription factors CBF/PEBP2 and c-Myb (CLEVERS and FERRIER 1998). Mutations that abolished binding of CBF/PEBP2 (LAUZURICA et al. 1997) or c-Myb (HERNANDEZ-MUNAIN et al. 1996) to δE3 in vitro also eliminated transcriptional activation by Eδ in transient transfection and abolished (LAUZURICA, ZHONG et al. 1997) or severely reduced (HERNANDEZ-MUNAIN, LAUZURICA et al. 1996) enhancer-dependent recombination of the TCRδ minilocus.

### 4.2.2 DNA Methylation

A growing body of evidence implicates DNA methylation as a potential regulator of locus-specific V(D)J recombination. Methylation diminished the frequency of rearrangement in transgenic or extrachromosomal substrates for V(D)J recombination (ENGLER et al. 1991; HSIEH and LIEBER 1992). Strikingly, demethylation of a stably transfected, methylated, κE-containing plasmid was found to be B-lineage-specific and enhancer-dependent (LICHTENSTEIN et al. 1994). Demethylation in B cells was induced by a compound, *cis*-acting element composed of the minimal κE, a MAR, and a sequence located downstream of Jκ5. More recently, monoallelic demethylation at the Igκ locus has been linked to the establishment of allelic exclusion (MOSTOSLAVSKY et al. 1998). In pro-B cells from *RAG*-$1^{-/-}$ mice, the κ locus is fully methylated, while in pre-B cells the κ locus is undermethylated. In a single $\kappa^+$ B cell, a fully methylated germline κ allele coexists with an undermethylated, productively rearranged κ allele. Optimal demethylation in vivo depends on the presence of iEκ and the 3′Eκ (MOSTOSLAVSKY et al. 1998).

Lineage-specific demethylation has also been observed at the Igμ locus. Eμ is flanked on either side by MARs. In a trangenic construct, the core Eμ and its flanking MARs were required for $V_H$ promoter activity in pre-B cells (FORRESTER et al. 1994). A transgene containing the Eμ core, Iμ, and MARs was demethylated in pre-B cells but not in non-lymphoid cells; demethylation was independent of the transcriptional state of the transgene but was markedly reduced in constructs lacking either MAR or the Iμ promoter (JENUWEIN et al. 1997).

### 4.2.3 Chromatin Structure

Genes undergoing V(D)J recombination are more sensitive to DNase-I digestion than transcriptionally inactive genes (see SLECKMAN et al. 1996 for review), consistent with the tenets of the accessibility hypothesis. Locus specificity may be more directly linked to chromatin structure than to transcription. When purified intact nuclei were examined for accessibility of specific loci to recombinase activity in vitro, lineage- and stage-specificity was maintained despite the absence of transcription (STANHOPE-BAKER et al. 1996). Specificity was lost upon purification of genomic DNA, indicating a dependence on noncovalent associations that maintain particular antigen-receptor gene segments accessible to cleavage by the RAG proteins. These results remain consistent, however, with roles for DNA methylation in directing changes in chromatin structure or in a parallel pathway for control of locus specificity.

### 4.2.4 Signaling Pathways that Affect Locus Accessibility

#### 4.2.4.1 BCR-Mediated Control of Stage- and Locus-Specific Rearrangement

In the B lineage, *RAG*-deficient mice exhibit a developmental arrest at the $B220^+CD43^+CD25^-$ stage; in the T lineage, a similar block is seen at the $CD4^-CD8^-CD44^{lo}CD25^+$ stage (Fig. 1). The blockade of B-cell development can be overcome in a stepwise manner by introduction of a transgenic Igμ chain (which takes cells to the $B220^+CD43^-CD25^+$ stage) or Igμ plus IgL (which yields $sIg^+$ B cells; see PAPAVASILIOU et al. 1997 for review). B-cell development is impaired at a similar stage by deficiency of membrane μ or the λ5 surrogate light chain (EHLICH et al. 1993), although in the latter case a substantial proportion of cells (about 5%) can bypass the block, perhaps because they are rescued by early rearrangement of the Igκ locus. Indeed, a combination of Igκ and Igμ transgenes rescued B-cell development in *RAG*-$1^{-/-}\lambda5^{-/-}$ mice, while Igμ alone could not (PAPAVASILIOU et al. 1996). Thus, a combination of Igμ and surrogate or conventional light chain is required for the normal proB/preB-I to preB-II transition.

The essential developmental signals generated by the Igμ-surrogate light-chain complex are mediated by its associated Igα/Igβ signal-transducing heterodimer. The arrest at the proB/preB-I to preB-II transition seen in *RAG*-deficient mice can be overcome by cross-linking of Igβ (NAGATA et al. 1997), indicating that the sIg-associated signaling complex can induce differentiation independent of effects on V(D)J recombination. Moreover, transgenic lymphocytes bearing a mutant μ chain that cannot associate with Igβ exhibit a block in maturation similar to that seen in *RAG*-deficient mice (IGLESIAS et al. 1993). Homozygous disruption of the gene encoding the Syk tyrosine kinase, which associates with Igα/Igβ and is activated upon BCR engagement, results in a profound block in B-cell development at the proB/preB-I to preB-II transition (CHENG et al. 1995). A recent study suggests that Igβ can exert locus-specific effects on V(D)J recombination (GONG and NUSSENZWEIG 1996). Mice deficient in Igβ exhibited a complete block in B-cell development at the $B220^+CD43^+$ stage; strikingly, a stage-specific block in IgH-chain rearrangement was observed, in that $D_HJ_H$ rearrangements were present, but $V_HD_HJ_H$ rearrangement was impaired (GONG and NUSSENZWEIG 1996). The latter observation is consistent with roles for Igβ in the activation of $V_H$ to $D_HJ_H$ joining or in the selection of productive $V_HD_HJ_H$ rearrangements.

#### 4.2.4.2 Regulation by T-Cell Accessory Molecules

The initial stages of TCR signaling are dependent on interaction of the tyrosine kinases Lck and Syk or ZAP-70 with each other and with the TCR (see LITTMAN and WEISS 1994 for review). Upon TCR engagement, Lck, through its association with the accessory molecules CD4 or CD8, is brought near to the TCR, where it participates in the recruitment of Syk or ZAP-70 to the TCR-associated CD3 complex and activates these kinases. As described earlier, functional TCRβ rearrangement blocks further β rearrangement (allelic exclusion) and permits differentiation from the CD4$^-$CD8$^-$ to CD4$^+$CD8$^+$ stage, at which TCRα rearrangement occurs. Cells that have recently undergone productive TCRβ rearrangement express an immature form of the TCR, in which the β chain is associated with pre-Tα; mice deficient for the TCRβ-associated surrogate light-chain pre-Tα exhibit impairment of T-cell development at the CD4$^-$CD8$^-$ to CD4$^+$CD8$^+$ transition (Fehling et al. 1995).

Developmental signals generated by pre-Tα/TCRβ are mediated by the associated CD3 chains and the accessory molecules CD4 or CD8. Indeed, signaling through the CD3 complex can drive development of β-chain-deficient T-cell progenitors past the CD4$^-$CD8$^-$ developmental block (Levelt et al. 1993). Mice deficient in either Syk or ZAP-70 show no defect in the formation of CD4$^+$CD8$^+$ thymocytes. In mice deficient in both kinases, however, CD4$^-$CD8$^-$ cells undergo β rearrangement but fail to proliferate and differentiate into CD4$^+$CD8$^+$ cells upon expression of pre-Tα/TCRβ. Thus Syk and ZAP-70 play overlapping roles in early thymocyte development (Cheng et al. 1997).

Manipulation of Lck activity in transgenic mice suggests a central role for this kinase in the signaling pathway by which the pre-TCR regulates rearrangement. First, overexpression of activated Lck in thymocytes interferes with $V_\beta$ rearrangement, but permits generation of normal numbers of CD4$^+$CD8$^+$ cells (Anderson et al. 1992). Second, a dominant negative Lck transgene permits β rearrangements to continue, despite expression of a transgenic β chain (Anderson et al. 1993). Third, thymocyte maturation was induced in *RAG*-2-deficient mice by introduction of a CD4 transgene; rescue of the maturation defect required both the ability of transgenic CD4 to bind Lck and the expression of MHC class II (Norment et al. 1997). Fourth, thymopoiesis in pre-Tα-deficient mice was rescued by treatment with anti-CD3 antibody or by introduction of a transgene encoding an active form of the Lck tyrosine kinase (Fehling et al. 1997). These observations support the hypothesis that Lck participates in the enforcement of allelic exclusion and the promotion of differentiation by the pre-Tα/TCRβ complex.

#### 4.2.4.3 Control of T-Cell Lineage Commitment by Notch

Although the mechanism by which a common T-cell precursor becomes an αβ or γδ T cell has not yet been completely resolved, compelling new studies indicate a role for the Notch signaling pathway (Robey and Fowlkes 1998). The murine protein Notch1 is a homolog of *Drosophila* Notch, a transmembrane receptor that delivers signals governing cell fate during embryogenesis. Expression of an activated form of Notch1 in murine thymocytes leads to both an increase in CD8-lineage T cells and a decrease in CD4-lineage T cells. Expression of activated Notch permits development of CD8-lineage thymocytes in the absence of class-I MHC proteins, but not in the absence of both class-I and class-II MHCs. Thus, Notch1 appears to participate in the CD4–CD8 lineage decision (Robey et al. 1996). More recently (Washburn et al. 1997), Notch1 has been implicated in the choice between αβ or γδ T-cell fates. T cells that are heterozygous for a mutant form of Notch1 were found to be biased toward development as γδ T cells, suggesting that reduced Notch1 activity favors the $\gamma\delta$ T-cell fate. Conversely, a constitutively active Notch1 mutant induced thymocytes bearing functional TCR$\gamma\delta$ gene rearrangements to adopt the $\alpha\beta$ T-cell fate. Based on these results, it has been proposed that commitment to the $\alpha\beta$ versus the $\gamma\delta$ lineage depends on the intensity of signals delivered by the Notch1 receptor (WASHBURN et al. 1997).

#### 4.2.4.4 Locus-Specific Control of V(D)J Recombination by IL-7

In mice deficient in IL-7 or the IL-7R, expansion of early lymphocytes is severely impaired. It has been reported that these mice exhibit a complete lack of the $\gamma\delta$ T-cell lineage and a selective impairment of $\gamma$ rearrangement, and these observations suggested that IL-7 regulates accessibility of the $\gamma$ locus to the recombinase (MAKI et al. 1996; CANDÉIAS et al. 1997). A recent study reexamined the relationship between IL-7 and $\gamma\delta$ T-cell development, with somewhat different findings (MALISSEN et al. 1997). In $\gamma$c-deficient mice, which lack functional IL-7R, $\gamma\delta$ T cells were absent in adult thymus but were present in fetal thymus, albeit in reduced numbers. Moreover, in $\gamma$c-deficient mice expressing a transgenic $\gamma\delta$ TCR, $\gamma\delta$ T cells were still greatly underrepresented, and these cells, unlike the corresponding IL-7R-expressing cells, failed to express Bcl-2. Thus, survival of committed or successfully rearranged $\gamma\delta$ T cells, rather than a failure of $\gamma$ rearrangement, may account for defective $\gamma\delta$ T-cell development in $\gamma$c-deficient thymi (MALISSEN et al. 1997).

A less equivocal relationship between IL-7 and locus accessibility has been suggested by studies of recombination at the IgH locus in IL-7R-deficient mice (CORCORAN et al. 1998). In these animals, $V_H$ to $D_HJ_H$ joining was suppressed; the severity of impairment increased strikingly with the distance of the $V_H$ segment from the D cluster. Germline transcripts from distal, unrearranged $V_H$ segments were specifically silenced, and developmentally regulated changes in chromatin structure of $V_H$ segments at the 5′ end of the IgH locus were absent. Interestingly, expression of the transcription factor Pax5/BSAP, which is essential for $V_H$ to $D_HJ_H$ rearrangement (but not $D_H$ to $J_H$ rearrangement; see above) is impaired in the IL-7R-deficient mice (CORCORAN et al. 1998). Taken together, these results provide evidence that an extrinsic regulatory signal can alter substrate accessibility to the recombinase and raise the possibility that IL-7 regulates the primary repertoire.

## 5 Concluding Remarks

The control of V(D)J recombination can be viewed in terms of a set of regulatory mechanisms which act in concert to establish specificity at several levels. Modes of control include modulation of *RAG* transcription, cell-cycle-dependent changes in the stability of RAG-2 protein, and alterations in the accessibility to the recombinase of individual gene segments. During development, these processes are superimposed to establish lineage-, stage-, and locus-specific patterns of V(D)J rearrangement. In the past 5 years, a number of extracellular signals that regulate V(D)J recombination, and the intracellular targets on which these signals act, have been identified. Nonetheless, our knowledge of the detailed signaling pathways that lead from the cell surface to the recombinase genes and antigen-receptor segments is at present primitive. Identification of the *cis*- and *trans*-acting regulators of *RAG* expression and locus accessibility will be essential to a future understanding of V(D)J recombination in its biological context.

# References

Agrawal A, Eastman QM, Schatz DG (1998) Transposition mediated by RAG1 and RAG2 and its implications for the evolution of the immune system. Nature 394:744–751

Alt FW, Oltz EM, Young F, Gorman J, Taccioli G, Chen J (1992) VDJ recombination. Immunol Today 13:306–314

Anderson SJ, Abraham KM, Nakayama T, Singer A, Perlmutter RM (1992) Inhibition of T-cell receptor β-chain gene rearrangement by overexpression of the non-receptor protein tyrosine kinase $p56^{lck}$. EMBO J 11:4877–4886

Anderson SJ, Levin SD, Perlmutter RM (1993) Protein tyrosine kinase $p56^{lck}$ controls allelic exclusion of T-cell receptor β-chain genes. Nature 365:552–554

Bain G, Maandag ECR, Izon DJ, Amsen D, Kruisbeek AM, Weintraub B, Krop I, Schlissel MS, Feeney AJ, vanRoon M, vanderValk M, teRiele HPJ, Berns A, Murre (1994) E2 A proteins are required for proper B cell development and initiation of immunoglobulin gene rearrangements. Cell 79:885–892

Billips LG, Nunez CA, Bertrand FE, Stankovic AK, Gartland GL, Burrows PD, Cooper MD (1995) Immunoglobulin recombinase gene activity is modulated reciprocally by interleukin 7 and CD19 in B cell progenitors. J Exp Med 182:973–982

Blackwell TK, Alt FW (1989) Mechanism and developmental program of immunoglobulin gene rearrangement in mammals. Ann Rev Gen 23:605–636

Bories J-C, Demengeot J, Davidson L, Alt FW (1996) Gene-targeted deletion and replacement mutations of the T-cell receptor β-chain enhancer: the role of enhancer elements in controlling V(D)J recombination accessibility. Proc Natl Acad Sci USA 93:7871–7876

Bories J-C, Willerford DM, Grevin D, Davidson L, Camus A, Martin P, Stehelin D, Alt FW (1995) Increased T-cell apoptosis and terminal B-cell differentiation induced by inactivation of the Ets-1 proto-oncogene. Nature 377:635–638

Bouvier G, Watrin F, Naspetti M, Verthuy C, Naquet P, Ferrier P (1996) Deletion of the mouse T-cell receptor β gene enhancer blocks αβ T-cell development. Proc Natl Acad Sci USA 93:7877–7881

Boyer PD, Diamond RA, Rothenberg EV (1989) Changes in the inducibility of IL-2 receptor alpha chain and T-cell receptor expression during thymocyte differentiation in the mouse. J Immunol 142: 4121–4130

Candéias S, Muegge K, Durum SK (1997) IL-7 receptor and V(D)J recombination: trophic versus mechanistic actions. Immunity 6:501–508

Candéias S, Peschon JJ, Muegge K, Durum SK (1997) Defective T-cell receptor γ gene rearrangement in interleukin-7 receptor knockout mice. Immunol Letters 57:9–14

Capone M, Watrin F, Fernex C, Horvat B, Krippl B, Wu L, Scollay R, Ferrier P (1993) TCRβ TCRα gene enhancers confer tissue- and stage-specificity on V(D)J recombination events. EMBO J 12: 4335–4346

Carlson LM, Oettinger MA, Schatz DG, Masteller EL, Hurley EA, McCormack WT, Baltimore D, Thompson CB (1991) Selective expression of RAG-2 in chicken B cells undergoing immunoglobulin gene conversion. Cell 64:201–208

Chen J, Young F, Bottaro A, Stewart V, Smith RK, Alt FW (1993) Mutations of the intronic IgH enhancer and its flanking sequences differentially affect accessibility of the $J_H$ locus. EMBO J 12: 4635–4645

Cheng AM, Negishi I, Anderson SJ, Chan AC, Bolen J, Loh DY, Pawson T (1997) The Syk and ZAP-70 SH2-containing tyrosine kinases are implicated in pre-T cell receptor signaling. Proc Natl Acad Sci USA 94:9797–9801

Cheng AM, Rowley RB, Pao W, Hayday A, Bolen JB, Pawson T (1995) Syk tyrosine kinase required for mouse viability and B cell development. Nature 378:303–306

Chun JJM, Schatz DG, Oettinger MA, Jaenisch R, Baltimore D (1991) The recombination activating gene-1 (RAG-1) transcript is present in the murine central nervous system. Cell 64:189–200

Clevers H, Ferrier P (1998) Transcriptional control during T-cell development. Curr Opin Immunol 10:166–171

Corcoran AE, Riddell A, Krooshoop D, Venkitaraman AR (1998) Impaired immunoglobulin gene rearrangement in mice lacking the IL-7 receptor. Nature 391:904–907

Davis MM, Bjorkman PJ (1988) T-cell antigen receptor genes and T-cell recognition. Nature 334:395–402

Decker D, Boyle NE, Koziol JA, Klinman NR (1991) The expression of the Ig H chain repertoire in developing bone marrow B lineage cells. J Immunol 146:350–361

Demengeot J, Oltz EM, Alt FW (1995) Promotion of V(D)J recombinational accessibility by the intronic E κ element: role of the κ B motif. Int Immunol 7:1995–2003

Desiderio S, Lin W-C, Li Z (1996) The cell cycle and V(D)J recombination. Curr Top Microbiol Immunol 217:45–59

Difilippantonio MJ, McMahan CJ, Eastman QM, Spanopoulou E, Schatz DG (1996) RAG1 mediates signal sequence recognition and recruitment of RAG2 in V(D)J recombination. Cell 87:253–262

Dudley EC, Petrie HT, Shah LM, Owen MJ, Hayday AC (1994) T cell receptor β chain gene rearrangement and selection during thymocyte development in adult mice. Immunity 1:83–93

Ehlich A, Schaal S, Gu H, Kitamura D, Muller W, Rajewsky K (1993) Immunoglobulin heavy and light chain genes rearrange independently at early stages of B cell development. Cell 72:695–704

Engler P, Haasch D, Pinkert CA, Doglio L, Glymour M, Brinster R, Storb U (1991) A strain-specific modifier on mouse chromosome 4 controls the methylation of independent transgenic loci. Cell 65:939–947

Fehling HJ, Iritani BM, Krotkova A, Forbush KA, Laplace C, Perlmutter RM, von Boehmer H (1997) Restoration of thymopoiesis in pTα$^{-/-}$ mice by anti-CD3ε antibody treatment or with transgenes encoding activated lck or tailless pTα. Immunity 6:703–714

Fehling HJ, Krotkova A, Saint-Ruf C, vonBoehmer H (1995) Crucial role of the pre-T-cell receptor alpha gene in development of alpha beta but not gamma delta T cells. Nature 375:795–798

Fernex C, Caspone M, Ferrier P (1995) The V(D)J recombinational and transcriptional activities of the immunoglobulin heavy-chain intronic enhancer can be mediated through distinct protein binding sites in a transgenic substrate. Mol Cell Biol 15:3217–3226

Ferrier P, Krippl B, Blackwell TK, Furley AJW, Suh H, Winoto A, Cook WD, Hood L, Costantini F, Alt FW (1990) Separate elements control DJ and VDJ rearrangement in a transgenic recombination substrate. EMBO J 9:117–125

Forrester WC, van Genderen C, Jenuwein T, Grosschedl R (1994) Dependence of enhancer-mediated transcription of the immunoglobulin μ gene on nuclear matrix attachment regions. Science 265: 1221–1225

Fuller K, Storb U (1997) Identification and characterization of the murine Rag1 promoter. Mol Immunol 12:939–954

Gaffney PM, Lund J, Miller JS (1998) FLT-3 ligand and marrow stroma-derived factors promote CD3γ, CD3δ, CD3ξ, and RAG-2 gene expression in primary human CD34$^{+}$LIN$^{-}$DR$^{-}$ marrow progenitors. Blood 91:1662–1670

Gay D, Saunders T, Camper S, Weigert M (1993) Receptor editing: an approach by autoreactive B cells to escape self tolerance. J Exp Med 177:999–1008

Godfrey DI, Kennedy J, Mombaerts P, Tonegawa S, Zlotnik A (1994) Onset of TCR-β rearrangement and role of TCR-β expression during CD3-CD4-CD8-thymocyte differentiation. J Immunol 152:4783–4792

Godfrey DI, Kennedy J, Suda T, Zlotnik A (1993) A developmental pathway involving four phenotypically and functionally distinct subsets of CD3$^{-}$CD4$^{-}$CD8$^{-}$ triple-negative adult mouse thymocytes defined by CD44 and CD25 expression. J Immunol 150:4244–4252

Gong S, Nussenzweig MC (1996) Regulation of an early developmental checkpoint in the B cell pathway by Igβ. Science 272:411–414

Gorman JR, van der Stoep N, Monroe R, Cogne M, Davidson L, Alt FW (1996) The Ig κ 3′ enhancer influences the ratio of Ig κ versus Ig λ B lymphocytes. Immunity 5:241–252

Grawunder U, West RB, Lieber MR (1998) Antigen receptor gene rearragement. Curr Opin Immunol 10:172–180

Groettrup M, Ungewiss K, Azogui O, Palacios R, Owen MJ, Hayday AC, von Boehmer H (1993) A novel disulfide-linked heterodimer on pre-T cells consists of the T cell receptor beta chain and a 33kd glycoprotein. Cell 75:283–294

Groettrup M, von Boehmer H (1993) A role for a pre-T-cell receptor in T-cell development. Immunol Today 14:610–614

Han S, Zheng B, Schatz DG, Spanopoulou E, Kelsoe G (1996) Neoteny in lymphocytes: RAG1 and RAG2 expression in germinal center B cells. Science 274:2094–2097

Hayashi R, Takemori T, Kodama M, Suzuki M, Tsuboi A, Nagawa F, Sakano H (1997) The PU1 binding site is a *cis*-element that regulates pro-B/pre-B specificity of Vκ-Jκ joining. J Immunol 159:4145–4149

Hempel WM, Stanhope-Baker P, Mathieu N, Huang F, Schlissel M, Ferrier P (1998) Enhancer control of V(D)J recombination at the TCR β locus: differential effects on DNA cleavage and joining. Genes Dev 12:2305–2317

Hernandez-Munain C, Lauzurica P, Krangel MS (1996) Regulation of T cell receptor δ gene rearrangement by c-Myb. J Exp Med 183:289–293

Hertz M, Kouskoff V, Nakamura T, Nemazee D (1998) V(D)J recombinase induction in splenic B lymphocytes is inhibited by antigen-receptor signalling. Nature 394:292–295

Hertz M, Nemazee D (1997) BCR ligation induces receptor editing in $IgM^+IgD^-$ bone marrow B cells in vitro. Immunity 6:429–436

Hikida M, Mori M, Takai T, Tomochika K, Hamatani K, Ohmori H (1996) Reexpression of RAG-1 and RAG-2 genes in activated mature mouse B cells. Science 274:2092–2094

Hikida M, Nakayama Y, Yamashita Y, Kumazawa Y, Nishikawa S-I, Ohmori H (1998) Expression of recombination activating genes in germinal center B cells: involvement of interleukin 7 (IL-7) and the IL-7 receptor. J Exp Med 188:365–372

Hikida M, Ohmori H (1998) Rearrangement of λ light chain genes in mature B cells in vitro and in vivo: function of reexpressed recombination-activating gene (RAG) products. J Exp Med 187:795–799

Hiom K, Gellert M (1997) A stable RAG1-RAG2-DNA complex that is active in V(D)J cleavage. Cell 88:65–72

Hiramatsu R, Akagi K, Matsuoka M, Sakumi K, Nakamura H, Kingsbury L, David C, Hardy RR, Yamamura K-I, Sakano H (1995) The 3′ enhancer region determines the B/T specificity and pro-B/pre-B specificity of immunoglobulin Vκ-Jκ joining. Cell 83:1113–1123

Hoffman ES, Passoni L, Crompton T, Leu TMJ, Schatz DG, Koff A, Owen MJ, Hayday AC (1996) Productive T-cell receptor β-chain gene rearrangement: coincident regulation of cell cycle and clonality during development in vivo. Genes Dev 10:948–9662

Hsieh CL, Lieber MR (1992) CpG methylated minichromosomes become inaccessible for V(D)J recombination after undergoing replication. EMBO J 11:315–325

Iglesias A, Nichogiannopoulou A, Williams GS, Flaswinkel H, Köhler G (1993) Early B cell development requires μ signaling. Eur J Immunol 23:2622–2630

Itoh N, Yasunaga M, Hirashima M, Yoshida O, Nishikawa S (1996) Role of IL-7 and KL in activating molecules controlling the G1/S transition of B precursor cells. Int Immunol 8:317–323

Jenuwein T, Forrester WC, Fernandez-Herrero LA, Laible G, Dull M, Grosschedl R (1997) Extension of chromatin accessibility by nuclear matrix attachment regions. Nature 385:269–272

Karasuyama H, Rolink A, Shinkai Y, Young F, Alt FW, Melchers F (1994) The expression of Vpre-B/lambda 5 surrogate light chain in early bone marrow precursor B cells of normal and B cell-deficient mutant mice. Cell 77:133–143

Kitagawa K, Mori K, Kishi H, Tagoh H, Nagata T, Kurioka H, Muraguchi A (1996) Chromatin structure and transcriptional regulation of human RAG-1 gene. Blood 88:3785–3791

Komori T, Okada A, Stewart V, Alt FW (1993) Lack of N regions in antigen receptor variable region genes of TdT-deficient lymphocytes. Science 261:1171–1175

Kurioka H, Kishi H, Isshiki H, Tagoh H, Mori K, Kitagawa T, Nagata T, Dohi K, Muraguchi A (1996) Isolation and characterization of TATA-less promoter for the human RAG-1 gene. Mol Immunol 33:1059–1066

Lauzurica P, Krangel MS (1994) Enhancer-dependent and -independent steps in the rearrangement of a human T cell receptor δ transgene. J Exp Med 179:43–55

Lauzurica P, Krangel MS (1994) Temporal and lineage-specific control of T cell receptor α/δ gene rearrangement by T cell receptor α and δ enhancers. J Exp Med 179:1913–1921

Lauzurica P, Zhong X-P, Krangel MS, Roberts JL (1997) Regulation of T cell receptor δ gene rearrangement by CBF/PEBP2. J Exp Med 185:1193–1201
Levelt CN, Mombaerts P, Iglesias A, Tonegawa S, Eichmann K (1993) Restoration of early thymocyte differentiation in T-cell receptor β-chain-deficient mutant mice by transmembrane signaling through CD3ε. Proc Natl Acad Sci USA 90:11401–11405
Lewis S (1994) The mechanism of V(D)J joining: lessons from molecular, immunological and comparative analyses. Adv Immunol 56:27–150
Li W, Swanson P, Desiderio S (1997) RAG-1- and RAG-2-dependent assembly of functional complexes with V(D)J recombination substrates in solution. Mol Cell Biol 17:6932–6939
Li Y-S, Hayakawa K, Hardy RR (1993) The regulated expression of B lineage associated genes during B cell differentiation in bone marrow and fetal liver. J Exp Med 178:951–960
Li Z, Dordai DI, Lee J, Desiderio S (1996) A conserved degradation signal regulates RAG-2 accumulation during cell division and links V(D)J recombination to the cell cycle. Immunity 5:575–589
Lichtenstein M, Keini G, Cedar H, Bergman Y (1994) B cell-specific demethylation: a novel role for the intronic κ chain enhancer sequence. Cell 76:913–923
Lin H, Grosschedl R (1995) Failure of B-cell differentiation in mice lacking the transcription factor EBF. Nature 376:263–267
Lin W-C, Desiderio S (1993) Regulation of V(D)J recombination activator protein RAG-2 by phosphorylation. Science 260:953–959
Lin W-C, Desiderio S (1994) Cell cycle regulation of V(D)J recombination activating protein RAG-2. Proc Natl Acad Sci USA 91:2733–2737
Lin W-C, Desiderio S (1995) V(D)J recombination and the cell cycle. Immunol Today 16:279–289
Littman D, Weiss A (1994) Signal transduction by lymphocyte antigen receptors. Cell 76:263–274
Maki K, Sunaga S, Ikuta K (1996) The V-J recombination of T cell receptor-γ genes is blocked in interleukin-7 receptor-deficient mice. J Exp Med 184:2423–2427
Malissen M, Pereira P, Gerber DJ, Malissen B, DiSanto JP (1997) The common cytokine receptor γ chain controls survival of γ/δ T cells. J Exp Med 186:1277–1285
Maser RS, Monsen KJ, Nelms BE, Petrini JH (1997) hMre11 and hRad50 nuclear foci are induced during the normal cellular response to DNA double-strand breaks. Mol Cell Biol 17:6087–6096
McBlane JF, vanGent DC, Ramsden DA, Romeo C, Cuomo CA, Gellert M, Oettinger MA (1995) Cleavage at a V(D)J recombination signal requires only RAG1 and RAG2 proteins and occurs in two steps. Cell 83:387–395
McMahan CJ, Fink PJ (1998) RAG reexpression and DNA recombination at T cell receptor loci in peripheral $CD4^+$ T cells. Immunity 9:637–647
Melamed D, Kench JA, Grabstein K, Rolink A, Nemazee D (1997) A functional B cell receptor transgene allows efficient IL-7-independent maturation of B cell precursors. J Immunol 159:1233–1239
Menetski JP, Gellert M (1990) V(D)J recombination activity in lymphoid cell lines is increased by agents that elevate cAMP. Proc Natl Acad Sci USA 87:9324–9328
Mostoslavsky R, Singh N, Kirillov A, Pelanda R, Cedar H, Chess A, Bergman Y (1998) κ chain monoallelic demethylation and the establishment of alleleic exclusion. Genes Dev 12:1801–1811
Muegge K, Vila MP, Durum SK (1993) Interleukin-7: A cofactor for V(D)J rearrangement of the T cell receptor β gene. Science 261:93–95
Muthusamy N, KB, Leiden J (1995) Defective activation and survival of T cells lacking the Ets-1 transcription factor. Nature 377:639–642
Nagata K, Nakamura T, Kitamura F, Kuramochi S, Taki S, Campbell KS, Karasuyama H (1997) The Ig alpha/Ig beta heterodimer on mu-negative proB cells is competent for transducing signals to induce early B cell differentiation. Immunity 7:559–570
Neale GAM, Fitzgerald TJ, Goorha RM (1992) Expression of the V(D)J recombinase gene RAG-1 is tightly regulated and involves both transcriptional and post-transcriptional control. Mol Immunol 29:1457–1466
Norment AM, Forbush KA, Nguyen N, Malissen M, Perlmutter RM (1997) Replacement of pre-T cell receptor signaling functions by the CD4 coreceptor. J Exp Med 185:121–130
O'Brien DP, Oltz EM, VanNess BG (1997) Coordinate transcription and V(D)J recombination of the kappa immunoglobulin light-chain locus: NF-kappaB-dependent and -independent pathways of activation. Mol Cell Biol 17:3477–3487
Oettinger MA, Schatz DG, Gorka C, Baltimore D (1990) RAG-1 and RAG-2, adjacent genes that synergistically activate V(D)J recombination. Science 248:1517–1523

Papavasiliou F, Casellas R, Suh H, Qin X-F, Besmer E, Pelanda R, Nemazee D, Rajewsky K, Nussenzweig M (1997) V(D)J recombination in mature B cells: a mechanism for altering antibody responses. Science 278:298–301
Papavasiliou F, Jankovic M, Gong S, Nussenzweig MC (1997) Control of immunoglobulin gene rearrangements in developing B cells. Curr Opin Immunol 9:233–238
Papavasiliou F, Jankovic M, Nussenzweig MC (1996) Surrogate or conventional light chains are required for membrane immunoglobulin mu to activate the precursor B cell transition. J Exp Med 184: 2025–2030
Reynisdottir I, Polyak K, Iavarone A, Massagué J (1995) Kip/Cip and Ink4 Cdk inhibitors cooperate to induce cell cycle arrest in response to TGF-β. Genes Dev 9:1831–1845
Roberts JL, Lauzurica P, Krangel MS (1997) Developmental regulation of V(D)J recombination by the core fragment of the T cell receptor α enhancer. J Exp Med 185:131–140
Robey E, Chang D, Itano A, Cado D, Alexander H, Lans D, Weinmaster G, Salmon P (1996) An activated form of Notch influences the choice between CD4 and CD8 T cell lineages. Cell 87:483–492
Robey E, Fowlkes BJ (1998) The αβ versus γδ T-cell lineage choice. Curr Opin Immunol 10:181–187
Rolink A, Grawunder U, Haasner D, Strasser A, Melchers F (1993) Immature surface Ig+ B cells can continue to rearrange κ and λ L chain gene loci. J Exp Med 178:1263–1270
Rolink A, Grawunder U, Winkler TH, Karasuyama H, Melchers F (1994) IL-2 receptor alpha chain (CD25,Tac) expression defines a crucial stage in pre-B cell development. Int Immunol 6:1257–1264
Rolink A, Kudo A, Karasuyama H, Kikuchi Y, Melchers F (1991) Long-term proliferating early pre B cell lines and clones with the potential to develop to surface Ig-positive, mitogen reactive B cells in vitro and in vivo. EMBO J 10:327–336
Rolink A, Streb M, Melchers F (1991) The κ/λ ratio in surface immunoglobulin molecules on B lymphocytes differentiating from DHJH-rearranged during pre-B cell clones in vitro. Eur J Immunol 11:2895–2898
Roth D, Zhu C, Gellert M (1993) Characterization of broken DNA molecules associated with V(D)J recombination. Proc Natl Acad Sci USA 90:10788–10792
Roth DB, Menetski JP, Nakajima PB, Bosma MJ, Gellert M (1992) V(D)J recombination: broken DNA molecules with covalently sealed (hairpin) coding ends in scid mouse thymocytes. Cell 70:983–991
Saintruf C, Ungewiss K, Groettrup M, Bruno L, Fehling HJ, Von Boehmer H (1994) Analysis and expression of a cloned pre-T cell receptor gene. Science 266:1208–1212
Schatz DG, Oettinger MA, Baltimore D (1989) The V(D)J recombination activating gene, RAG-1. Cell 59:1035–1048
Schlissel M, Constantinescu A, Morrow T, Baxter M, Peng A (1993) Double-strand signal sequence breaks in V(D)J recombination are blunt, 5′-phosphorylated, RAG-dependent, and cell cycle regulated. Genes Dev 32:2520–2532
Schlissel MS, Voronova A, Baltimore D (1991) Helix loop helix transcription factor E47 activates germline immunoglobulin heavy-chain gene transcription and rearrangement in a pre-T cell line. Genes Dev 5:1367–1376
Serwe M, Sablitzky F (1993) V(D)J recombination in B cells is impaired but not blocked by targeted deletion of the immunoglobulin heavy chain intron enhancer. EMBO J 12:2321–2327
Shinkai Y, Koyasu S, Nakayama K, Murphy KM, Loh DY, Reinherz EL, Alt FW (1993) Restoration of T cell development in RAG-2-deficient mice by functional TCR transgenes. Science 259:822–825
Sleckman BP, Bardon CG, Ferrini R, Davidson L, Alt FW (1997) Function of the TCRα enhancer in αβ and γδ T cells. Immunity 7:505–515
Sleckman BP, Gorman JR, Alt FW (1996) Accessibility control of antigen-receptor variable-region gene assembly: role of cis-acting elements. Ann Rev Immunol 14:459–481
Spanopoulou E, Roman CAJ, Corcoran LM, Schlissel MS, Silver DP, Nemazee D, Nussenzweig MC, Shinton SA, Hardy RR, Baltimore D (1994) Functional immunoglobulin transgenes guide ordered B-cell differentiation in RAG-1-deficient mice. Genes Dev 8:1030–1042
Spanopoulou E, Zaitseva F, Wang F-H, Santagata S, Baltimore D, Panayotou G (1996) The homeodomain region of Rag-1 reveals the parallel mechanisms of bacterial and V(D)J recombination. Cell 87:263–276
Stanhope-Baker P, Hudson KM, Shaffer AL, Constantinescu A, Schlissel MS (1996) Cell type-specific chromatin structure determines the targeting of V(D)J recombinase activity in vitro. Cell 85:887–89
Swanson PC, Desiderio S (1998) V(D)J recombination signal recognition: distinct, overlapping DNA-protein contacts in complexes containing RAG1 with and without RAG2. Immunity 9:115–125

Tagoh H, Kishi H, Okumura A, Kitagawa T, Nagata T, Mori K, Muraguchi A (1996) Induction of recombination activating gene expression in a human lymphoid progenitor cell line: requirement of two separate signals from stromal cells and cytokines. Blood 88:4463–4473

Takahama Y, Letterio JJ, Suzuki H, Farr AG, Singer A (1994) Early progression of thymocytes along the CD4/CD8 developmental pathway is regulated by a subset of thymic epithelial cells expressing transforming growth factor β. J Exp Med 179:1495–1506

Takata M, Sasaki MS, Sonoda E, Morrison C, Hashimoto M, Utsumi H, Yamaguchi-Iwai Y, Shinohara A, Takeda S (1998) Homologous recombination and non-homologous end-joining pathways of DNA double-strand break repair have overlapping roles in the maintenance of chromosomal integrity in vertebrate cells. EMBO J 17:5497–5508

ten Boekel E, Melchers F, Rolink A (1995) The status of Ig loci rearrangements in single cells from different stages of B cell development. Int Immunol 7:1013–1019

Tiegs SL, Russell DM, Nemazee D (1993) Receptor editing in self-reactive bone marrow B cells. J Exp Med 177:1009–1020

Tsukada S, Saffran DC, Rawlings DJ, Parolini O, Allen RC, Klisak I, Sparkes RS, Kubagawa H, Mohandas T, Quan S, Belmont JW, Cooper MD, Conley ME, Witte ON (1993) Deficient expression of a B cell cytoplasmic tyrosine kinase in human X-linked agammaglobulinemia. Cell 72:279–290

Urbanek P, Wang Z-Q, Fetka I, Wagner EF, Busslinger M (1994) Complete block of early B cell differentiation and altered patterning of the posterior midbrain in mice lacking Pax5/BSAP. Cell 79:901–912

van der Stoep N, Gorman JR, Alt FW (1998) Reevaluation of 3′Ekappa function in stage- and lineage-specific rearrangement and somatic hypermutation. Immunity 8:743–750

van Gent DC, McBlane JF, Ramsden DA, Sadofsky MJ, Hesse JE, Gellert M (1995) Initiation of V(D)J recombination in a cell-free system. Cell 81:925–934

van Gent DC, Mizuuchi K, Gellert M (1996) Similarities between initiation of V(D)J recombination and retroviral integration. Science 271:1592–1594

Verbeek S, Izon D, Hofhuis F, Robanus-Maandag E, te Riele T, van de Wetering M, Oosterwegel M, Wilson A, MacDonald HR, Clevers H (1995) An HMG-box-containing T-cell factor required for thymocyte differentiation. Nature 374:70–74

Vetrie D, Vorechovsky I, Sideras P, Holland J, Davies A, Flinter F, Hammarstrom L, Kinnon C, Levinsky R, Bobrow M, Smith CIE, Bentley DR (1993) The gene involved in X-linked agammaglobulinemia is a member of the src family of protein-tyrosine kinases. Nature 361:226–233

Washburn T, Schweighoffer E, Gridley T, Chang D, Fowlkes BJ, Cado D, Robey E (1997) Notch activity influences the αβ versus γδ T cell lineage decision. Cell 88:833–843

Wilson A, Held W, MacDonald HR (1994) Two waves of recombinase gene expression in developing thymocytes. J Exp Med 179:1355–1360

Wolf ML, Weng WK, Stiegelbauer KT, Shah N, LeBien TW (1993) Functional effect of IL-7-enhanced CD19 expression on human B cell precursors. J Immunol 151:138–148

Xu Y, Davidson L, Alt FW, Baltimore D (1996) Deletion of the Ig κ light chain intronic enhancer/matrix attachment region impairs but does not abolish VkJ k rearrangement. Immunity 4:377–385

Zhuang Y, Soriano P, Weintraub H (1994) The helix-loop-helix gene E2 A is required for B cell formation. Cell 79:875–884

# B-Cell-Receptor-Dependent Positive and Negative Selection in Immature B Cells

D. Nemazee[1], V. Kouskoff[2], M. Hertz[3], J. Lang[4], D. Melamed[5], K. Pape[2], and M. Retter[6]



## 1 Early Stages in B-Cell Development

In a pro-B cell, the variable portions of the immunoglobulin (Ig) heavy (H)-chain genes are assembled in a quasi-random V(D)J-recombination process beginning with D-to-J joining and followed by V-to-DJ joining (reviewed in Tonegawa 1983). Presumably because of the error-prone nature of this gene assembly, the production of functional IgM H-chain protein (μ-chain) is monitored and selected on the basis of its ability to assemble with the "surrogate light-chain" components λ5 and VpreB (Karasuyama et al. 1996; Rajewsky 1996). The μ-chain/surrogate light-chain complex associates with the B-cell-specific signal transducers Ig-α and Ig-β through the μ-chain's transmembrane tail, forming a pre-B receptor (pre-B-cell receptor, preBCR; Karasuyama et al. 1996; Reth and Wienands 1997). It is this assembled preBCR that transmits a specific differentiation signal with four

[1] The Scripps Research Institute, 10550 North Torrey Pines Rd., IMM-29, La Jolla, CA 92037, USA
[2] The National Jewish Medical and Research Center, 1400 Jackson Street, Denver, CO 80206, USA
[3] M & E Biotech. A/S, kogle Alle 6, DK-2970 Hørsholm, Denmark
[4] Barbara Davis Center for Childhood Diabetes, 4200 East Ninth Avenue, Box B-140 Denver, CO 80262, USA
[5] Bruce Rappoport School of Medicine, Department of Immunology, Bat-Galin, Haifa 31905, Israel
[6] Corixa Corporation, 1124 Columbia Street, Suite 200, Seattle, WA 98104, USA

important and interrelated outcomes: (1) B-cell developmental progression, (2) downregulation of V(D)J recombination, (3) continued cell survival, and (4) changes in growth-factor sensitivity. This model is supported by data indicating that a complete, membrane form of μH-chain prevents assembly of a second H-chain gene (Nussenzweig et al. 1987; Storb 1987; Gu et al. 1991; Ehlich et al. 1994) and is required for B-cell development (Reichman-Fried et al. 1990; Kitamura et al. 1991; Spanopoulou et al. 1994; Young et al. 1994a). This differentiation is defined and monitored by a number of well-characterized surface markers and characteristic RNA expression patterns, as well as by the ability of cells to proliferate to bone-marrow-stromal cell-derived cytokines (Hardy et al. 1991; Li et al. 1993; Rolink et al. 1994; Grawunder et al. 1995).

There is some evidence that the signals involved in these steps may be separable. An incompletely formed DJ-Cμ gene sometimes has a protein product that can associate with the surrogate light (L) chains (Tsubata et al. 1991a) and turn off V(D)J recombination, but that cannot foster continued development (Gu et al. 1991; Ehlich et al. 1994). A DJ-Cμ transgene has similar effects (Tornberg et al. 1998). Similarly, in certain Ig-transgenic mice, H-chain proteins have been identified that can apparently mediate allelic exclusion but not full developmental progression (Roth et al. 1993). This may be related to the ability of the H-chains to associate with Ig-α/β (Cronin et al. 1998). Importantly, the downregulation of V(D)J recombination stimulated by the preBCR is very efficient, such that cells with one productive H allele almost always downregulate recombination before a second such rearrangement occurs on the other allele, yielding a significant fraction of B cells with VDJ assembly on only one allele (Alt et al. 1984; Ehlich et al. 1994).

Consistent with the notion that an assembled preBCR is necessary for allelic exclusion signals, in the absence of the surrogate L-chain component λ5 preB cells bearing two functional μ chains can be found at early developmental stages (Ehlich et al. 1994; Loffert et al. 1996), but allelic exclusion is restored at later stages by unknown mechanisms (Kitamura et al. 1992). Conversely, cells with two non-productive H alleles or cells that cannot rearrange Ig genes owing to a defect in recombinase fail to mature and die rapidly (Bosma et al. 1983; Reichman-Fried et al. 1990; Hardy et al. 1991; Kitamura et al. 1991; Ehlich et al. 1994; Spanopoulou et al. 1994). This checkpoint in development is also associated with low-level expression of survival genes (Li et al. 1993). Enforced expression of Bcl-2 or Bcl-xl was shown to be capable of rescuing B cells with aberrant H-chain rearrangements from apoptosis, but failed to allow their developmental progression (Fang et al. 1996; Young et al. 1997b).

As reviewed elsewhere in this volume, it is widely assumed that preBCR signaling is similar to BCR signaling in more mature B cells. One clear difference is that there is some doubt about the precise cellular distribution and subcellular localization of preB receptors in vivo. Biochemical evidence for the presence of a cell-surface-expressed preBCR present on primary bone-marrow preB cells is comparatively weak (Karasuyama et al. 1994; Karasuyama et al. 1996), but transformed preB cell lines express higher levels of preBCR (Pillai and Baltimore

1988; MISENER et al. 1991; TSUBATA et al. 1992b; LASSOUED et al. 1993; ROLINK et al. 1993). In addition, the stages of development at which the preBCR is seen on the cell surface has been disputed and may differ between mouse and human (CHERAYIL and PILLAI 1991; LASSOUED et al. 1993). In any case, the evidence for the current model for the role of the preB receptor is almost exclusively based on genetic rather than biochemical data.

Gene-ablation experiments have shown that signaling molecules important in BCR-mediated activation of mature B cells, such as syk, btk, and Ig-α/β, are required for normal B-cell development (KHAN et al. 1995; GONG and NUSSENZWEIG 1996; TORRES et al. 1996; TYBULEWICZ 1998). Experiments with chimeric μ-chain/Ig-α/β constructs bearing intact or tyrosine-to-phenylalanine-mutated immunoreceptor tyrosine-based activation motifs (ITAMs) showed that intact Ig-β or Ig-α ITAMs were critical and sufficient for developmental progression and downregulation of H-loci rearrangements in recombinase-deficient preB cells (PAPAVASILIOU et al. 1995, 1995a). Studies have indicated that this preB-cell signal also upregulates L-chain gene rearrangements (TSUBATA et al. 1992b; SPANOPOULOU et al. 1994; PAPAVASILIOU et al. 1995; CONSTANTINESCU and SCHLISSEL 1997; TARAKHOVSKY 1997). Mice deficient in V(D)J recombination, Ig-α, Ig-β, or the membrane form of μ manifest a complete block in B-cell development, whereas mice deficient in syk, btk, or λ5 have impaired B-cell development; nevertheless, some cells do develop. In humans, unlike mice, natural mutations in btk completely ablate B-cell production at the preB stage (TARAKHOVSKY 1997; SATTERTHWAITE et al. 1998). In mice deficient in V(D)J recombination or the membrane form of μ, the block in B-cell development can be relieved by the introduction of Ig transgenes (REICHMAN-FRIED et al. 1990; HARDY et al. 1991; SPANOPOULOU et al. 1994; YOUNG et al. 1994a), demonstrating that Ig is limiting for preBCR signaling. However, the requirement of preBCR expression for development can be bypassed in certain experimental contexts, such as by cross-linking Ig-α/β (NAGATA et al. 1997) or by treatment with interleukin (IL)-4 and anti-CD40 antibodies (ROLINK et al. 1996).

It is not clear whether the preBCR signals upon interaction with an endogenous or exogenous ligand or whether it has no ligand at all. Mice deficient in λ5 are severely impaired in their ability to promote B-cell development (KITAMURA et al. 1992; ROLINK et al. 1993), but are somewhat "leaky", because the production of authentic Ig L chains allows their rescue (PAPAVASILIOU et al. 1996b; PELANDA et al. 1996). These data suggest that surrogate L chain serves as a chaperone for μ chain. In apparent confirmation of this hypothesis, it was recently shown that a transgene encoding a truncated μH chain lacking the V and CH1 domains and, therefore, unable to pair with either surrogate or authentic L chains, could nevertheless promote B-cell development (SHAFFER and SCHLISSEL 1997), as could a human heavy-chain disease protein transgene (CORCOS et al. 1995). In addition, certain μ chains that fail to foster developmental progression (DECKER et al. 1991; YE et al. 1995) are unable to associate with surrogate L chain because of their CDR3 structure (KEYNA et al. 1995; YE et al. 1996). Collectively, it now appears that the ability of μ chain to assemble into a preBCR complex is a major differentiative signal, driving preB cell proliferation and subsequent differentiation, at least in conventional, B-2-lineage cells. Recent data indicate that preB cells of the B-1 lineage may respond to preBCR signaling in a different way than B-2 preB cells (WASSERMAN et al. 1998). Interestingly, at a comparable stage in T-cell development, T-cell receptor (TCR)β expression is monitored by heterodimerization and signaling through the surrogate TCRα chain, preTα (VON BOEHMER and FEHLING 1997). In this context, the extracellular domains of the preTCR are dispensable for function, i.e. developmental progression (IRVING et al. 1998), but cell-surface localization of preTα is probably required (O'SHEA et al. 1997).

It is convenient in discussing the early stages in mouse B-cell development to refer to the nomenclature of Richard Hardy (HARDY et al. 1991), who extensively characterized the cell-surface markers defining several intermediate stages in the transition from B-lineage-committed precursors (fraction A) to mature, recirculating B cells (fraction F). In fraction C (CD45R+/CD43+/BP-1+/HSA intermediate), cells undergo V-to-(D)J assembly on the heavy-chain loci, generating some cells with functional μ chains. Through the putative signaling mechanisms mentioned above, cells with functional μ chains are rapidly recruited into the fraction C′ compartment, which expresses a higher surface level of HSA (heat-stable antigen, CD24) than fraction C and undergoes very rapid proliferation. This expansion is limited by availability of preBCR signaling and lymphokines such as IL-7 (SUDO et al. 1989; GRABSTEIN et al. 1993). These cells also downregulate V(D)J recombinase (HARDY et al. 1991; GRAWUNDER et al. 1995). After as many as six cell divisions (DECKER et al. 1991), fraction C′ cells stop proliferating and differentiate to a small preB cell phenotype with distinctive cell-surface markers (fraction D: CD43-/sIg-/CD45R intermedate). These cells now express and redirect V(D)J-recombinase activity to the Ig L-chain gene loci (RETH et al. 1985; CONSTANTINESCU and SCHLISSEL 1997), allowing the formation of complete antigen receptors, although some L-chain-gene rearrangement is independent of H-chain expression (reviewed by GORMAN and ALT 1998). Little is known about the mechanisms controlling DNA accessibility to recombinase (reviewed by GORMAN and ALT 1998), but inhibition of V(D)J recombination on the remaining non-productive H-chain allele, which often has a D-to-J join, must be uncoupled from H-chain-gene expression in order to maintain H-chain allelic exclusion while allowing L-chain-gene rearrangement.

## 2 Emergence of Surface sIgM+ Immature B Cells

As pre-B cells undergo functional L-chain gene rearrangements, L chains are produced and assemble with μ chains, eventually giving rise to the assembly on the cell surface of complete IgM antigen-receptor molecules. This is the earliest point in B-cell development at which complete surface IgM (sIgM) is expressed on the plasma membrane. This stage is subject to both positive selection, which promotes further differentiation, and negative selection imposed by immune tolerance. How both signals are transmitted through a single antigen receptor is an interesting problem with parallels to that of T-cell development in CD4+8+ thymocytes.

## 3 Positive Selection as a Process of Continued Developmental Progression

The first known function of sIgM is to promote both further B-cell development and the cessation of Ig L-chain rearrangements. This function has been inferred from the observations of allelic exclusion of L-chain loci and the ability of L-chain transgenes to suppress endogenous L-chain expression (RITCHIE et al. 1984; STORB 1987). Expression of cell-surface IgM is correlated with downregulation of recombinase-activator gene (*RAG*) expression (LI et al. 1993), and IgM transgenes suppress *RAG* expression (TIEGS et al. 1993; MELAMED et al. 1997). This developmental progression is sometimes referred to as "positive selection" (MELCHERS et al. 1995), by analogy to the positive-selection process occurring during thymic development of T cells, in which T cells with "appropriate" receptors are allowed to survive and mature, whereas cells lacking receptors or bearing receptors that fail to have some affinity for major histocompatibility complex (MHC) molecules fail to mature and are left to die. In T cells, as in B cells, the process of antigen-receptor-driven positive selection is likely directed by common intracellular signaling pathways and is associated with the cessation of V(D)J recombination and *RAG* transcription (BORGULYA et al. 1992). These events can be partly replicated in vitro by stimulation of the TCR with antibodies (TURKA et al. 1991) or antigen in the appropriate context (SHAO et al. 1997). In positive selection of both T and B cells, the cells acquire new cell-surface molecules and eventually migrate to the periphery. In the case of T cells, positive selection clearly requires specific, but very weak, engagement of the TCR with (auto)antigenic ligands borne by self MHC molecules, a requirement that is believed to be critical for shaping the T-cell repertoire to recognize foreign antigens only in the context of self MHC molecules (JAMESON et al. 1995).

Shortly after sIgM expression, B cells clearly experience an important developmental checkpoint that is blocked in the absence of L-chain expression (REICHMAN-FRIED et al. 1990; SPANOPOULOU et al. 1994; YOUNG et al. 1994a). Appropriate L-chain expression allows both maturation of surface phenotypes and migration of cells to the peripheral lymphoid tissues. This developmental step is blocked by ablation of BCR-signaling molecules and is associated with upregulation of Bcl2 expression (LI et al. 1993; MERINO et al. 1994). In B cells, a requirement for extracellular BCR ligands in positive selection is not established experimentally, nor justified in theory. Nevertheless, it cannot yet be ruled out that self-antigens promote the survival or differentiation of some B cells. This could occur in several ways: myriad weak interactions of the antigen-combining site with self antigens or idiotopes could appropriately stimulate the BCR. Alternatively, a monomorphic self-antigen could bind to constant regions of sIg, signaling positive selection. It has been suggested that, in rabbits, the CD5 molecule can associate with *VH* framework determinants and select a particular *VH* family (POSPISIL et al. 1996). Another possibility is that cells with impaired BCR signaling might be rescued by the compensatory effects of antigen binding to the BCR. Examples of such an antigen-mediated rescue in signaling-deficient or receptor-insufficient mice have been presented (ANDERSSON et al. 1995; CYSTER et al. 1996), and similar effects are likely to be seen in comparable experimental situations. However, it is rather likely that in normal B cells the basal or "tonic" signaling of an unligated BCR may itself provide sufficient signal to promote B-cell development. If true, this would represent a major difference between development in T cells and B cells.

Why should positive selection in B cells require no specific ligand? There are several possible explanations. First, unlike T cells, there is no evidence that B-cell responses are restricted by association to another molecule such as MHC; therefore, the *raison d'être* of antigen-specific positive selection is apparently missing. Second, unlike the TCR, which is probably monomeric on the cell surface, the assembled BCR is generated as a covalent dimer of signaling tails because of the disulfide bonds that hold together IgH chains. This predicts that the mere assembly of the BCR dimerizes the Ig-α/β transducers and presumably drives a certain level of transactivation of associated kinases. (Speaking against this interpretation is the fact that, even in "monomeric" form, the CD3 complex contains a predimerized zeta sub-unit along with other ITAM-containing sub-units.) Third, preliminary experiments with transgenic mice expressing truncated sIgM Fc molecules lacking Fab portions (and therefore antigen-combining sites) suggest that positive selection can occur in the absence of specific antigen binding (SHAFFER and SCHLISSEL 1997). However, there is increasing evidence that, in mouse mutants with impaired BCR signaling, B cells with autoreactive receptors selectively survive (CYSTER et al. 1996), suggesting that weak expression or signaling defects in somatic cells might sometimes favor autoreactive BCRs. This rescue of autoreactive cells has potentially important implications for the etiology of autoimmunity.

## 4 Tolerance-Mediated Negative Selection as a Developmental Block

Immune tolerance in emerging sIgM cells appears to occur through a combination of developmental arrest, receptor editing, and apoptosis (GAY et al. 1993; HARTLEY et al. 1993; RADIC et al. 1993; MONROE 1996; MELAMED and NEMAZEE 1997; MELAMED et al. 1998). In the negative-selection process, sIgM signaling must carry a signal distinct from or antagonistic to that which governs positive selection, indicating that, similar to the situation in developing $TCR\alpha\beta^+CD4^+CD8^+$ thymocytes, the quantity or quality of the antigen-receptor signal for negative selection is distinct from that of positive selection. While this distinction is not known in molecular detail, in practice it appears that cross-linking of the BCR by antibodies or antigens is usually sufficient to generate a negative signal. In any case, the initial outcome of negative selection is the prevention of developmental progression. However, unlike the situation in thymocyte development, this developmental arrest is not associated with rapid cell death, as cells with autoreactive receptors die no faster than cells that have not yet managed to generate an IgM receptor (MELAMED and NEMAZEE 1997; MELAMED et al. 1998). Importantly, this developmental arrest is reversible; if the cell should alter its autoreactive receptor and replace it with an "innocent" one, the cell maintains the potential for full development. Thus, the developmental checkpoints in proB-to-preB and preB-to-B development represent not only culling steps, but also provide opportunities for receptor-gene repair and improvement.

L-chain genes that generate autoreactive receptors or that are out of frame can be repaired by nested rearrangements at the same locus, as often occurs in the κ loci, or can be destroyed by aberrant nested rearrangements followed by new rearrangements at the other L-chain loci. It is not unusual for developing B cells to undergo two, three, or more L-chain gene rearrangements (LEWIS et al. 1982; VAN NESS et al. 1982; COLECLOUGH 1983; FEDDERSEN and VAN NESS 1985; SHAPIRO and WEIGERT 1987; HARADA and YAMAGISHI 1991; LOWENSON and CLARKE 1991). This ongoing rearrangement is capable of causing receptor editing, because the organization of the L-chain loci permits nested rearrangements that can inactivate the cell's active L-chain gene and replace it with another. Also, because of the error-prone nature of such rearrangements, the original L-chain VJ element may be removed prior to replacement with a functional one. This means that the normal order of developmental progression may be reversed, with sIgM+ cells giving rise to preB cells that may or may not be rescued subsequently by functional L-chain-gene assembly. This has been observed experimentally in Ig-gene-targeted replacement mice (CHEN et al. 1997; PELANDA et al. 1997). In any case, the time frame available to a particular cell for undergoing developmental arrest and receptor editing is not unlimited, lasting only a day or two in the mouse bone marrow, and cells failing to generate functional, non-autoreactive receptors will die.

## 5 The Role of Cell Survival

Regardless of the ligand dependence of B-cell positive selection, it is evident that a necessary checkpoint must be cleared in order for continued B-cell development to proceed and for extended survival and peripheralization to occur. Because κ L-chain genes generally rearrange prior to λ loci in B-cell development (ARAKAWA et al. 1996), the longer a cell takes to generate a functional receptor, the more likely it is to rearrange and express λ genes. Interestingly, in Bcl-2-overexpressing mice, the time window for receptor editing appears to be extended and the fraction of B cells expressing λ increases from 6% to 15% (LANG et al. 1997). This result and the fact that most λ+ B cells have undergone only a single rearrangement attempt at a λ locus suggest that the time window of receptor editing is sufficiently short that many cells do not have time to exhaust their L-chain loci before dying. The ability of Bcl-2 overexpression to extend survival and to promote continued L-chain gene recombination has been demonstrated in vitro (ROLINK et al. 1993). Cells failing to advance in development turn over rapidly and fail to populate the peripheral immune tissues. Overexpression of death-inhibitory molecules, such as Bcl-2, also allows extended survival and peripheralization of preB cells (HARTLEY et al. 1993; YOUNG et al. 1997b) and extended survival of B cells, including the plasma cell, at several stages of development (STRASSER et al. 1991). The related death inhibitor bcl-xl has also been shown to have the capacity to suppress programmed cell death associated with tolerance and to promote receptor editing (FANG et al. 1998).

## 6 Positive Selection into the Long-Lived Peripheral Pool

In contrast to the receptor-editing mode of negative selection, B cells that are somewhat more advanced in development and have undergone positive selection, also known as the "transitional" B-cell population, are known to be exquisitely sensitive to very rapid antigen-induced apoptosis, at least in experimental models of tolerance (ALLMAN et al. 1992; CARSETTI et al. 1995; MELAMED et al. 1998). The vast majority of these newly formed, recombinase non-expressing cells migrate to the periphery and turn over rapidly in vivo (with a life span of a few days). A residual, small subset of these transitional cells enters the long-lived B-cell pool and acquires a lifespan of many weeks (FÖRSTER and RAJEWSKY 1990). It is unclear to what extent the cell loss at this transition is due to negative selection or to the lack of some affirmative signaling. Sequencing studies clearly demonstrated that the pool of cells that is long lived has an overrepresentation of some *VH* gene sequences and underuses other genes that are abundantly rearranged in preB cells, indicating that recruitment to the long-lived pool is in some measure dependent on specificity (GU et al. 1991). Furthermore, mature, resting B cells that are induced to lose Ig expression by cre/lox recombination die rapidly, indicating that continued signaling through the BCR is required for B-cell survival (LAM et al. 1997). However, this signaling may represent a basal or tonic signal and may not need interaction with specific ligands. Furthermore, in BCR-transgenic mice, in which virtually all B cells have the same specificity, transgene-expressing cells are present in both short-lived and long-lived pools (FULCHER and BASTEN 1994), arguing that specificity alone does not determine cell fate, but that other resources, such as space constraints, limit the survival possibilities of new B cells produced in the bone marrow.

## 7 Antigen-Receptor Signaling in Positive and Negative Selection

As discussed elsewhere in this volume, BCR signaling is complex and incompletely understood. Most biochemical information about signaling has been gleaned from the analysis of cell lines or primary splenic B cells (Scott 1993; Reth and Wienands 1997; Healy and Goodnow 1998). Consequently, our understanding about the biochemistry of BCR signaling in less mature cells, such as newly formed bone-marrow B cells, is based mainly on extrapolation from studies of BCR signaling in these other contexts. Furthermore, studies on less mature cells have so far focused on responses to strong stimuli, such as anti-BCR antibodies (Igarashi et al. 1994; Monroe 1996). In many ways, the most useful conceptual framework for understanding how B-cell positive and negative selection can be mediated by the same receptor comes from recent data concerning T-cell-receptor signaling.

In T cells, "differential signaling" refers to the finding that "weak" antigenic stimuli can stimulate biological responses that are qualitatively distinct from strong stimuli. This ability of the TCR to signal in more than one way is thought to provide a biochemical explanation for the distinction between positive and negative selection in the thymus and activation versus partial activation or antagonism in peripheral T cells. For example, T-cell lines stimulated with "strong" ligands proliferate and produce abundant cytokines, whereas weaker ligands often stimulate cytokine production or anergy in the absence of proliferation (Evavold and Allen 1991; Sloan-Lancaster and Allen 1996) or can even inhibit responses to full agonists (Sette et al. 1994). Altered peptide ligands are mutants of strong agonist peptides that often have the ability to mediate partial activation. Several studies suggest that such ligands are remarkably efficient at the process of thymic positive selection (Hogquist and Bevan 1996). Thus, differential signaling is believed to play a role in both immature and mature T-cell responses, and the biochemical basis of this differential activation has been extensively studied.

Recent studies have focused on the role of TCR-mediated phosphorylation events on the CD3zeta chain, which can differ in partial activation compared with full activation (Kersh et al. 1998; Madrenas et al. 1995). These differences have more severe consequences on subsequent recruitment and phosphorylation of downstream effectors, such as ZAP-70, which can only be activated upon bidentate binding of its tandem SH2 domains to doubly phosphorylated ITAMs, but may be recruited without activation to partly phosphorylated CD3 components (Madrenas et al. 1995). CD3zeta chains exist in T cells as dimers (Alberola-Ila et al. 1997). Since each CD3zeta monomer contains three ITAMs, each of which contains two tyrosine phosphorylation sites [Yxx(L/I)xxxxxxYxx(L/I)], and since the different sites have numerous sequence differences, recent studies have focused on the degree to which particular motifs are phosphorylated during full or partial activation. Surprisingly, there appears to be a rather sophisticated ordering to the phosphorylations that depends upon the nature of the stimulus.

According to a recent study by Allen and colleagues (Kersh et al. 1998), upon stimulation with strong ligands, all six sites in CD3zeta are phosphorylated, and mutational studies demonstrated that the phosphorylation of certain tyrosines are contingent on the presence of others, allowing a determination of the order of phosphorylation. Most importantly, partial agonists or antagonists caused partial and abortive phosphorylation of particular sites on CD3zeta, providing a potentially diverse array of covalently modified CD3zeta forms (and other CD3 components) with diverse possible biological potentials. In the BCR complex, the signal transducers Ig-α and Ig-β each contain an ITAM, each of which contains a pair of tyrosine-phosphorylation sites. It is tempting to speculate that the pattern of phosphorylation and downstream consequences may differ in situations of positive versus negative selection. Although Ig-α and Ig-β are functionally redundant in certain situations (Papavasiliou et al. 1995; Teh and Neuberger 1997), they probably have non-overlapping functions. Using a platelet-derived growth factor receptor-Ig-α/β chimeric molecule approach, Clark's group has proposed that Ig-α and Ig-β can have both distinct and cooperative functions, and that phosphorylations of these molecules themselves can be synergistic (Luisiri et al. 1996). It will be important to assess the functional significance of these differences and to relate them to the biological selection processes. Another insight from T-cell studies with probable applicability to B-cell processes is that, while antigen receptor activation is required for both positive and negative selection, the Ras/mitogen-activated protein kinase biochemical pathways are not required for negative selection, but may be required for positive selection (Swan et al. 1995; Alberola-Ila et al. 1996; Shao et al. 1997). Initial studies using dominant negative Ras (H-rasN17) or activated *raf* transgenes suggest that this is also the case for B cells (Iritani et al. 1997).

## 8 Summary

This review touches on only a small part of the complex biology of B cells, but serves to illustrate the point that the antigen receptor is the most important of many cell-surface receptors affecting cell-fate decisions. Receptor expression is necessary, but not sufficient, for cell survival. It is also essential that a B cell's antigen-receptor specificity be appropriate for its environment. The need to balance reactivity with self tolerance has resulted in an intricate feedback control (affected by both the recombinase and cell survival) that regulates independent selection events at the level of the receptor and the cell.

## References

Alberola-Ila J, Hogquist KA, Swan KA, Bevan MJ, Perlmutter RM (1996) Positive and negative selection invoke distinct signaling pathways. J Exp Med 184:9–18

Alberola-Ila J, Takaki S, Kerner JD, Perlmutter RM (1997) Differential signaling by lymphocyte antigen receptors. Annu Rev Immunol 15:125–154

Allman DM, Ferguson SE, Cancro MP (1992) Peripheral B cell maturation. I. Immature peripheral B cells in adults are heat-stable antigenhi and exhibit unique signaling characteristics. J Immunol 149:2533–2540
Alt FW, Yancopoulos GD, Blackwell TK, Wood C, Thomas E, Boss M, Coffman R, Rosenberg N, Tonegawa S, Baltimore D (1984) Ordered rearrangement of immunoglobulin heavy chain variable region segments. EMBO J 3:1209–1219
Andersson J, Melchers F, Rolink A (1995) Stimulation by T cell independent antigens can relieve the arrest of differentiation of immature auto-reactive B cells in the bone marrow. Scand J Immunol 42:21–33
Arakawa H, Shimizu T, Takeda S (1996) Re-evaluation of the probabilities for productive arrangements on the kappa and lambda loci. Int Immunol 8:91–99
Borgulya P, Kishi H, Uematsu Y, von Boehmer H (1992) Exclusion and inclusion of alpha and beta T cell receptor alleles. Cell 69:529–537
Bosma GC, Custer RP, Bosma MJ (1983) A severe combined immunodeficiency mutation in the mouse. Nature 301:527–530
Carsetti R, Kohler G, Lamers MC (1995) Transitional B cells are the target of negative selection in the B cell compartment. J Exp Med 181:2129–2140
Chen C, Prak EL, Weigert M (1997) Editing disease-associated autoantibodies. Immunity 6:97–105
Cherayil BJ, Pillai S (1991) The omega/lambda 5 surrogate immunoglobulin light chain is expressed on the surface of transitional B lymphocytes in murine bone marrow. J Exp Med 173:111–116
Coleclough C (1983) Chance, necessity and antibody gene dynamics. Nature 303:23–26
Constantinescu A, Schlissel MS (1997) Changes in locus-specific V(D)J recombinase activity induced by immunoglobulin gene products during B cell development. J Exp Med 185:609–620
Corcos D, Dunda O, Butor C, Cesbron JY, Lores P, Bucchini D, Jami J (1995) Pre-B-cell development in the absence of lambda 5 in transgenic mice expressing a heavy-chain disease protein. Current Biology 5:1140–1148
Cronin FE, Jiang M, Abbas AK, Grupp SA (1998) Role of mu heavy chain in B cell development. I. Blocked B cell maturation but complete allelic exclusion in the absence of Ig alpha/beta. J Immunol 161:252–259
Cyster JG, Healy JI, Kishihara K, Mak TW, Thomas ML, Goodnow CC (1996) Regulation of B-lymphocyte negative and positive selection by tyrosine phosphatase CD45. Nature 381:325–328
Decker DJ, Boyle NE, Klinman NR (1991) Predominance of nonproductive rearrangements of VH81X gene segments evidences a dependence of B cell clonal maturation on the structure of nascent H chains. J Immunol 147:1406–1411
Ehlich A, Martin V, Muller W, Rajewsky K (1994) Analysis of the B-cell progenitor compartment at the level of single cells. Curr Biol 4:573–583
Evavold BD, Allen PM (1991) Separation of IL-4 production from Th cell proliferation by an altered T cell receptor ligand. Science 252:1308–1310
Fang W, Mueller DL, Pennell CA, Rivard JJ, Li YS, Hardy RR, Schlissel MS, Behrens TW (1996) Frequent aberrant immunoglobulin gene rearrangements in pro-B cells revealed by a bcl-xL transgene. Immunity 4:291–299
Fang W, Weintraub BC, Dunlap B, Garside P, Pape KA, Jenkins MK, Goodnow CC, Mueller DL, Behrens TW (1998) Self-reactive B lymphocytes overexpressing Bcl-xL escape negative selection and are tolerized by clonal anergy and receptor editing. Immunity 9:35–45
Feddersen RM, Van Ness BG (1985) Double recombination of a single immunoglobulin kappa-chain allele: implications for the mechanism of rearrangement. Proc Natl Acad Sci USA 82:4793–4797
Förster I, Rajewsky K (1990) The bulk of the peripheral B-cell pool in mice is stable and not rapidly renewed from the bone marrow. Proc Natl Acad Sci USA 87:4781–4784
Fulcher DA, Basten A (1994) Reduced life span of anergic self-reactive B cells in a double-transgenic model. J Exp Med 179:125–134
Gay D, Saunders T, Camper S, Weigert M (1993) Receptor editing: an approach by autoreactive B cells to escape tolerance. J Exp Med 177:999–1008
Gong S, Nussenzweig MC (1996) Regulation of an early developmental checkpoint in the B cell pathway by Ig beta. Science 272:411–414
Gorman JR, Alt FW (1998) Regulation of immunoglobulin light chain isotype expression. Adv Immunol 69:113–181
Grabstein KH, Waldschmidt TJ, Finkelman FD, Hess BW, Alpert AR, Boiani NE, Namen AE, Morrissey PJ (1993) Inhibition of murine B and T lymphopoiesis in vivo by an anti-interleukin 7 monoclonal antibody. J Exp Med 178:257–264

Grawunder U, Leu TM, Schatz DG, Werner A, Rolink AG, Melchers F, Winkler TH (1995) Down-regulation of RAG1 and RAG2 gene expression in preB cells after functional immunoglobulin heavy chain rearrangement. Immunity 3:601–608
Gu H, Kitamura D, Rajewsky K (1991) B cell development regulated by gene rearrangement: arrest of maturation by membrane-bound D mu protein and selection of DH element reading frames. Cell 65:47–54
Gu H, Tarlinton D, Muller W, Rajewsky K, Forster I (1991) Most peripheral B cells in mice are ligand selected. J Exp Med 173:1357–1371
Harada K, Yamagishi H (1991) Lack of feedback inhibition of V kappa gene rearrangement by productively rearranged alleles. J Exp Med 173:409–415
Hardy RR, Carmack CE, Shinton SA, Kemp JD, Hayakawa K (1991) Resolution and characterization of pro-B and pre-pro-B cell stages in normal mouse bone marrow. J Exp Med 173:1213–1225
Hartley SB, Cooke MP, Fulcher DA, Harris AW, Cory S, Basten A, Goodnow CC (1993) Elimination of self-reactive B lymphocytes proceeds in two stages: arrested development and cell death. Cell 72: 325–335
Healy JI, Goodnow CC (1998) Positive versus negative signaling by lymphocyte antigen receptors. Annu Rev Immunol 16:645–670
Hogquist KA, Bevan MJ (1996) The nature of the peptide/MHC ligand involved in positive selection. Semin Immunol 8:63–68
Igarashi H, Kuwahara K, Nomura J, Matsuda A, Kikuchi K, Inui S, Sakaguchi N (1994) B cell Ag receptor mediates different types of signals in the protein kinase activity between immature B cell and mature B cell. J Immunol 153:2381–2393
Iritani BM, Forbush KA, Farrar MA, Perlmutter RM (1997) Control of B cell development by Ras-mediated activation of Raf. EMBO J 16:7019–7031
Irving BA, Alt FW, Killeen N (1998) Thymocyte development in the absence of pre-T cell receptor extracellular immunoglobulin domains. Science 280:905–908
Jameson SC, Hogquist KA, Bevan MJ (1995) Positive selection of thymocytes. Annu Rev Immunol 13:93–126
Karasuyama H, Rolink A, Melchers F (1996) Surrogate light chain in B cell development. Adv Immunol 63:1–41
Karasuyama H, Rolink A, Shinkai Y, Young F, Alt FW, Melchers F (1994) The expression of Vpre-B/lambda 5 surrogate light chain in early bone marrow precursor B cells of normal and B cell-deficient mutant mice. Cell 77:133–143
Kersh EN, Shaw AS, Allen PM (1998) Fidelity of T cell activation through multistep T cell receptor zeta phosphorylation. Science 281:572–575
Keyna U, Applequist SE, Jongstra J, Beck-Engeser GB, Jack HM (1995) Ig mu heavy chains with VH81X variable regions do not associate with lambda 5. Ann NY Acad Sci 764:39–42
Khan WN, Alt FW, Gerstein RM, Malynn BA, Larsson I, Rathbun G, Davidson L, Muller S, Kantor AB, Herzenberg LA (1995) Defective B cell development and function in Btk-deficient mice. Immunity 3:283–299
Kitamura D, Kudo A, Schaal S, Muller W, Melchers F, Rajewsky K (1992) A critical role of lambda 5 protein in B cell development. Cell 69:823–831
Kitamura D, Roes J, Kuhn R, Rajewsky K (1991) A B cell-deficient mouse by targeted disruption of the membrane exon of the immunoglobulin mu chain gene. Nature 350:423–426
Lam KP, Kuhn R, Rajewsky K (1997) In vivo ablation of surface immunoglobulin on mature B cells by inducible gene targeting results in rapid cell death. Cell 90:1073–1083
Lang J, Arnold B, Hammerling G, Harris AW, Korsmeyer S, Russell D, Strasser A, Nemazee D (1997) Enforced Bcl-2 expression inhibits antigen-mediated clonal elimination of peripheral B cells in an antigen dose-dependent manner and promotes receptor editing in autoreactive, immature B cells. J Exp Med 186:1513–1522
Lassoued K, Nunez CA, Billips L, Kubagawa H, Monteiro RC, LeBlen TW, Cooper MD (1993) Expression of surrogate light chain receptors is restricted to a late stage in pre-B cell differentiation. Cell 73:73–86
Lewis S, Rosenberg N, Alt F, Baltimore D (1982) Continuing kappa-gene rearrangement in a cell line transformed by Abelson murine leukemia virus. Cell 30:807–816
Li YS, Hayakawa K, Hardy RR (1993) The regulated expression of B lineage associated genes during B cell differentiation in bone marrow and fetal liver. J Exp Med 178:951–960

Loffert D, Ehlich A, Muller W, Rajewsky K (1996) Surrogate light chain expression is required to establish immunoglobulin heavy chain allelic exclusion during early B cell development. Immunity 4:133–144
Lowenson JD, Clarke S (1991) Structural elements affecting the recognition of L-isoaspartyl residues by the L-isoaspartyl/D-aspartyl protein methyltransferase. Implications for the repair hypothesis. J Biol Chem 266:19396–19406
Luisiri P, Lee YJ, Eisfelder BJ, Clark MR (1996) Cooperativity and segregation of function within the Ig-alpha/beta heterodimer of the B cell antigen receptor complex. J Biol Chem 271:5158–5163
Madrenas J, Wange RL, Wang JL, Isakov N, Samelson LE, Germain RN (1995) Zeta phosphorylation without ZAP-70 activation induced by TCR antagonists or partial agonists. Science 267:515–518
Melamed D, Benschop RJ, Cambier JC, Nemazee D (1998) Developmental regulation of B lymphocyte immune tolerance compartmentalizes clonal selection from receptor selection. Cell 92:173–182
Melamed D, Kench JA, Grabstein K, Rolink A, Nemazee D (1997) A functional B cell receptor transgene allows efficient IL-7-independent maturation of B cell precursors. J Immunol 159:1233–1239
Melamed D, Nemazee D (1997) Self-antigen does not accelerate immature B cell apoptosis, but stimulates receptor editing as a consequence of developmental arrest. Proc Natl Acad Sci USA 94:9267–9272
Melchers F, Rolink A, Grawunder U, Winkler TH, Karasuyama H, Ghia P, Andersson J (1995) Positive and negative selection events during B lymphopoiesis. Curr Opin Immunol 7:214–227
Merino R, Ding L, Veis DJ, Korsmeyer SJ, Nunez G (1994) Developmental regulation of the Bcl-2 protein and susceptibility to cell death in B lymphocytes. EMBO J 13:683–691
Misener V, Downey GP, Jongstra J (1991) The immunoglobulin light chain related protein lambda 5 is expressed on the surface of mouse pre-B cell lines and can function as a signal transducing molecule. Int Immunol 3:1129–1136
Monroe JG (1996) Tolerance sensitivity of immature-stage B cells: can developmentally regulated B cell antigen receptor (BCR) signal transduction play a role? J Immunol 156:2657–2660
Nagata K, Nakamura T, Kitamura F, Kuramochi S, Taki S, Campbell KS, Karasuyama H (1997) The Ig alpha/Igbeta heterodimer on mu-negative proB cells is competent for transducing signals to induce early B cell differentiation. Immunity 7:559–570
Nussenzweig MC, Shaw AC, Sinn E, Danner DB, Holmes KL, Morse HC, Leder P (1987) Allelic exclusion in transgenic mice that express the membrane form of immunoglobulin mu. Science 236:816–819
O'Shea CC, Thornell AP, Rosewell IR, Hayes B, Owen MJ (1997) Exit of the pre-TCR from the ER/cis-Golgi is necessary for signaling differentiation, proliferation, and allelic exclusion in immature thymocytes. Immunity 7:591–599
Papavasiliou F, Jankovic M, Nussenzweig MC (1996b) Surrogate or conventional light chains are required for membrane immunoglobulin mu to activate the precursor B cell transition. J Exp Med 184:2025–2030
Papavasiliou F, Jankovic M, Suh H, Nussenzweig MC (1995a) The cytoplasmic domains of immunoglobulin (Ig) alpha and Ig beta can independently induce the precursor B cell transition and allelic exclusion. J Exp Med 182:1389–1394
Papavasiliou F, Misulovin Z, Suh H, Nussenzweig MC (1995) The role of Ig beta in precursor B cell transition and allelic exclusion. Science 268:408–411
Pelanda R, Schaal S, Torres RM, Rajewsky K (1996) A prematurely expressed Ig(kappa) transgene, but not V(kappa)J(kappa) gene segment targeted into the Ig(kappa) locus, can rescue B cell development in lambda5-deficient mice. Immunity 5:229–239
Pelanda R, Schwers S, Sonoda E, Torres RM, Nemazee D, Rajewsky K (1997) Receptor editing in a transgenic mouse model: site, efficiency, and role in B cell tolerance and antibody diversification. Immunity 7:765–775
Pillai S, Baltimore D (1988) The omega and iota surrogate immunoglobulin light chains. Curr Top Microbiol Immunol 137:136–139
Pospisil R, Fitts MG, Mage RG (1996) CD5 is a potential selecting ligand for B cell surface immunoglobulin framework region sequences. J Exp Med 184:1279–1284
Radic MZ, Erikson J, Litwin S, Weigert M (1993) B lymphocytes may escape tolerance by revising their antigen receptors. J Exp Med 177:1165–1173
Rajewsky K (1996) Clonal selection and learning in the antibody system. Nature 381:751–758
Reichman-Fried M, Hardy RR, Bosma MJ (1990) Development of B-lineage cells in the bone marrow of scid/scid mice following the introduction of functionally rearranged immunoglobulin transgenes. Proc Natl Acad Sci USA 87:2730–2734

Reth M, Wienands J (1997) Initiation and processing of signals from the B cell antigen receptor. Annu Rev Immunol 15:453–479

Reth MG, Ammirati P, Jackson S, Alt FW (1985) Regulated progression of a cultured pre-B-cell line to the B-cell stage. Nature 317:353–355

Ritchie KA, Brinster RL, Storb U (1984) Allelic exclusion and control of endogenous immunoglobulin gene rearrangement in kappa transgenic mice. Nature 312:517–520

Rolink A, Grawunder U, Winkler TH, Karasuyama H, Melchers F (1994) IL-2 receptor alpha chain (CD25, TAC) expression defines a crucial stage in pre-B cell development. Int Immunol 6:1257–1264

Rolink A, Karasuyama H, Grawunder U, Haasner D, Kudo A, Melchers F (1993) B cell development in mice with a defective lambda 5 gene. Eur J Immunol 23:1284–1288

Rolink A, Melchers F, Andersson J (1996) The SCID but not the RAG-2 gene product is required for S mu-S epsilon heavy chain class switching. Immunity 5:319–330

Roth PE, Doglio L, Manz JT, Kim JY, Lo D, Storb U (1993) Immunoglobulin gamma 2b transgenes inhibit heavy chain gene rearrangement, but cannot promote B cell development. J Exp Med 178:2007–2021

Satterthwaite AB, Li Z, Witte ON (1998) Btk function in B cell development and response. Semin Immunol 10:309–316

Scott DW (1993) Analysis of B cell tolerance in vitro. Adv Immunol 54:393–425

Sette A, Alexander J, Ruppert J, Snoke K, Franco A, Ishioka G, Grey HM (1994) Antigen analogs/MHC complexes as specific T cell receptor antagonists. Annu Rev Immunol 12:413–431

Shaffer AL, Schlissel MS (1997) A truncated heavy chain protein relieves the requirement for surrogate light chains in early B cell development. J Immunol 159:1265–1275

Shao H, Kono DH, Chen LY, Rubin EM, Kaye J (1997) Induction of the early growth response (Egr) family of transcription factors during thymic selection. J Exp Med 185:731–744

Shao H, Rubin EM, Chen LY, Kaye J (1997) A role for Ras signaling in coreceptor regulation during differentiation of a double-positive thymocyte cell line. J Immunol 159:5773–5776

Shapiro MA, Weigert M (1987) How immunoglobulin V kappa genes rearrange. J Immunol 139: 3834–3839

Sloan-Lancaster J, Allen PM (1996) Altered peptide ligand-induced partial T cell activation: molecular mechanisms and role in T cell biology. Annu Rev Immunol 14:1–27

Spanopoulou E, Roman CA, Corcoran LM, Schlissel MS, Silver DP, Nemazee D, Nussenzweig MC, Shinton SA, Hardy RR, Baltimore D (1994) Functional immunoglobulin transgenes guide ordered B-cell differentiation in Rag-1-deficient mice. Gen Dev 8:1030–1042

Storb U (1987) Transgenic mice with immunoglobulin genes. Annu Rev Immunol 5:151–174

Strasser A, Whittingham S, Vaux DL, Bath ML, Adams JM, Cory S, Harris AW (1991) Enforced BCL2 expression in B-lymphoid cells prolongs antibody responses and elicits autoimmune disease. Proc Natl Acad Sci USA 88:8661–8665

Sudo T, Ito M, Ogawa Y, Iizuka M, Kodama H, Kunisada T, Hayashi S, Ogawa M, Sakai K, Nishikawa S (1989) Interleukin 7 production and function in stromal cell-dependent B cell development. J Exp Med 170:333–338

Swan KA, Alberola-Ila J, Gross JA, Appleby MW, Forbush KA, Thomas JF, Perlmutter RM (1995) Involvement of p21ras distinguishes positive and negative selection in thymocytes. EMBO J 14: 276–285

Tarakhovsky A (1997) Xid and Xid-like immunodeficiencies from a signaling point of view. Curr Opin Immunol 9:319–323

Teh YM, Neuberger MS (1997) The immunoglobulin (Ig)alpha and Igbeta cytoplasmic domains are independently sufficient to signal B cell maturation and activation in transgenic mice. J Exp Med 185:1753–1758

Tiegs SL, Russell DM, Nemazee D (1993) Receptor editing in self-reactive bone marrow B cells. J Exp Med 177:1009–1020

Tonegawa S (1983) Somatic generation of antibody diversity. Nature 302:575–581

Tornberg UC, Bergqvist I, Haury M, Holmberg D (1998) Regulation of B lymphocyte development by the truncated immunoglobulin heavy chain protein Dmu. J Exp Med 187:703–709

Torres RM, Flaswinkel H, Reth M, Rajewsky K (1996) Aberrant B cell development and immune response in mice with a compromised BCR complex. Science 272:1804–1808

Tsubata T, Tsubata R, Reth M (1991a) Cell surface expression of the short immunoglobulin mu chain (D mu protein) in murine pre-B cells is differently regulated from that of the intact mu chain. Eur J Immunol 21:1359–1363

Tsubata T, Tsubata R, Reth M (1992b) Crosslinking of the cell surface immunoglobulin (mu-surrogate light chains complex) on pre-B cells induces activation of V gene rearrangements at the immunoglobulin kappa locus. Int Immunol 4:637–641
Turka LA, Schatz DG, Oettinger MA, Chun JJ, Gorka C, Lee K, McCormack WT, Thompson CB (1991) Thymocyte expression of RAG-1 and RAG-2: termination by T cell receptor cross-linking. Science 253:778–781
Tybulewicz VL (1998) Analysis of antigen receptor signalling using mouse gene targeting. Curr Opin Cell Biol 10:195–204
Van Ness BG, Coleclough C, Perry RP, Weigert M (1982) DNA between variable and joining gene segments of immunoglobulin kappa light chain is frequently retained in cells that rearrange the kappa locus. Proc Natl Acad Sci USA 79:262–266
von Boehmer H, Fehling HJ (1997) Structure and function of the pre-T cell receptor. Annu Rev Immunol 15:433–452
Wasserman R, Li YS, Shinton SA, Carmack CE, Manser T, Wiest DL, Hayakawa K, Hardy RR (1998) A novel mechanism for B cell repertoire maturation based on response by B cell precursors to pre-B receptor assembly. J Exp Med 187:259–264
Ye J, McCray SK, Clarke SH (1995) The majority of murine VH12-expressing B cells are excluded from the peripheral repertoire in adults. Eur J Immunol 25:2511–2521
Ye J, McCray SK, Clarke SH (1996) The transition of pre-BI to pre-BII cells is dependent on the VH structure of the mu/surrogate L chain receptor. EMBO J 15:1524–1533
Young F, Ardman B, Shinkai Y, Lansford R, Blackwell TK, Mendelsohn M, Rolink A, Melchers F, Alt FW (1994a) Influence of immunoglobulin heavy- and light-chain expression on B-cell differentiation. Gen Dev 8:1043–1057
Young F, Mizoguchi E, Bhan AK, Alt FW (1997b) Constitutive Bcl-2 expression during immunoglobulin heavy chain-promoted B cell differentiation expands novel precursor B cells. Immunity 6:23–33

# CD40-CD154 Interactions in B-Cell Signaling

D.M. CALDERHEAD, Y. KOSAKA, E.M. MANNING, and R.J. NOELLE



## 1 Introduction

CD40 and its ligand CD154, [also called gp39, CD40L, tumor necrosis factor-related activation protein, and T-cell/B-cell activation molecule (TBAM)] have been shown to play a pivotal, non-redundant role in the regulation of both humoral

Department of Microbiology, Dartmouth Medical School, Lebanon, NH 03781, USA and cell-mediated immunity (Fig. 1). The critical role that this ligand-receptor pair plays is underscored by the fact that blockade of this ligand-receptor pair globally inhibits tetanus-diphtheria (TD) humoral immunity and many components of cell-mediated immunity. Furthermore, the development and progression of a wide spectrum of autoimmune diseases (experimental allergic encephalomyelitis, collagen-induced arthritis, lupus, etc.), as well as the responses to transplantation antigens [graft-vs-host disease (GVHD), cardiac allografts, skin allografts], are halted upon therapeutic intervention with αCD154, as reviewed in Foy (1996). Over the past few years, substantial strides have been made in identifying CD40-associated molecules that may mediate CD40 signaling.

## 1.1 CD40 and CD154 (CD40L) Structure and Expression

CD40 is a 50-kDa, integral membrane protein and a member of the tumor-necrosis-factor receptor (TNF-R) family of receptors, which include TNF-R1, TNF-R2, p75 nerve-growth-factor receptor (NGFR), CD30, CD27, Fas, TNF-related apoptosis-inducing ligand, TNF-related activation-induced cytokine and others. The extracellular domain consists of 193 amino acids, which include 22 cysteine residues, a feature common to members of this family. The transmembrane and cytoplasmic

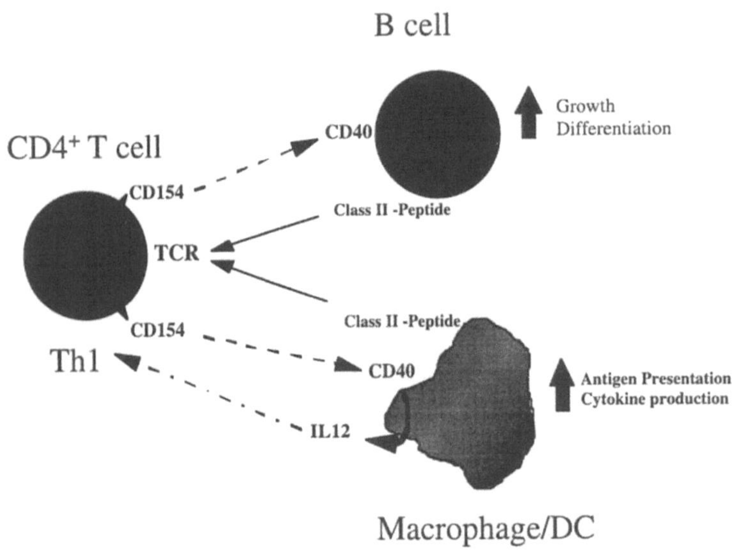


**Fig. 1.** CD40–CD154 regulates humoral and cell-mediated immunity. Helper T cells provide CD154 as a necessary stimulus for the development of humoral and cellular based immune responses. Cognate interactions between helper T cells and B cells result in ligation of CD40 on the B cell, triggering effector phase functions such as clonal expansion, differentiation, and secretion of high affinity immunoglobulins. Similarly, helper T cells ligate CD40 during cognate interactions with professional APCs, which are induced to greater functional presentation accompanied by the elicitation of cytokines which are vital for cellular responses

domains are 22 and 62 amino acids in length, respectively. Although related to other TNF-R members through homology in their extracellular domains, there is limited homology in the cytoplasmic domains of the various members. However, the limited homologies in the cytoplasmic domain do demonstrate a consensus amongst TNF-R family members for binding common downstream signaling molecules (TNF-R-associated factors, TRAFs; see below).

CD154 is a 33- to 36-kDa type-II integral membrane glycoprotein that is a member of the TNF family of ligands. The extracellular domain of the molecule is composed of a 61-residue membrane-proximal spacer region and a 149-residue TNF-homologous, receptor-binding domain, which is C-terminal to the spacer region. CD154 has a short cytoplasmic domain consisting of 22 amino acids. Modeling studies have suggested that CD154 exists as a trimer on the surface of activated T cells and, as such, can engage multiple CD40 receptors, which facilitates clustering and triggering (BAJORATH 1995).

The tissue distribution of CD40 was originally thought restricted to B cells, dendritic cells and basal epithelial cells (CLARK 1990). More recently, however, it has been recognized that the tissue distribution of CD40 is much broader and that it may play an important functional role in a wide spectrum of cells and tissues (Table 1). Unlike the receptor, CD154 expression is far more restricted. For the most part, prominent expression of CD154 is found only on activated $CD4^+$ helper T cells. The expression is rapidly induced by T-cell-receptor (TCR) engagement and is transient. The expression is seen on Th0, Th1 and Th2 clones, some $CD8^+$ T-cell

**Table 1.** Expression and function of CD40

| | Function | Reference |
|---|---|---|
| CD40 expression | | |
| Hematopoietic | | |
| B cells | Growth and differentiation, Ig class switching | STAMENKOVIC 1989; FOY 1996 |
| Macrophages | Cytokine production, effector functions | ALDERSON 1993 |
| Dendritic cells, Langerhan's cells, mast cells, basophils | Cytokine production, antigen presentation | BANCHEREAU 1995; CAUX 1994 |
| Non-hematopoietic | | |
| Fibroblasts | Cytokine production | |
| Endothelial cells | Adhesion, cytokine production | KARMANN 1995; HOLLENBAUGH 1995 |
| Microglia, astrocytes, keratinocytes | Cytokines, antigen presentation | ALOISI 1998; TAN 1998 |
| CD154 expression | | |
| Hematopoietic | | |
| CD4+ T cells | Provides help | ROY 1993; ARMITAGE 1992 |
| CD8+ T cells | Provides help | CRONIN 1995 |
| Activated B cells | Provides help | GRAMMER 1995 |
| Mast cells, basophils | Provides help | GAUCHAT 1993 |
| Eosinophils | | GAUCHAT 1995 |
| Non-hematopoietic | | |
| None | | |

clones and weakly on activated B cells. In addition to lymphocytes, CD154 expression has been reported on activated mast cells, eosinophils and basophils.

## 1.2 CD40–CD154 Interactions in the Regulation of Humoral Immunity

Insights into the function of the ligand for CD154 in the regulation of humoral immunity have been provided by studies in CD154 – (Xu 1994; Renshaw 1994) and CD40 – (Kawabe 1994) – deficient mice and in mice treated with αCD154 (Foy 1993, 1994; Van den Eertwegh 1993). For the most part, all of these systems agree that CD154–CD40 interactions are essential for secondary immune responses to TD antigens and in the formation of germinal centers (GC). Studies using αCD154 (Foy 1994) or a soluble form of CD40 (Gray 1994) to inhibit the in vivo function of CD154 show that the generation of memory B cells is also dependent on this ligand-receptor pair. Studies in humans with the immunodeficiency hyper-immunoglobulin (Ig)M syndrome (Allen 1993; Aruffo 1993; DiSanto 1993; Korthauer 1993) provided proof that mutations in the CD154 gene rendered humans profoundly immunosuppressed.

CD40 is a potent activator of B-cell growth and differentiation, as reviewed by Banchereau (1994). CD40 induces a series of decisive events in B cells which turn out to be pivotal points in the life history of the mature B cell during the time it is interacting with its cognate antigen-specific helper T cell. Using recombinant forms of CD154 or agonistic αCD40 monoclonal antibodies, it has been shown that engagement of CD40 and cytokines [most notably interleukin (IL)4)] can induce profound B-cell proliferation. The importance of this ligand in regulating B-cell clonal expansion in vivo has also been confirmed (Garside 1998). In addition to the induction of B-cell growth, the activation of B cells with CD40 and cytokines has been demonstrated to be extremely efficient in inducing Ig isotype switching (Banchereau 1994). One of the earliest events that commits a B cell to isotype switching is the expression of sterile transcripts from the specific locus to be rearranged (Jumper 1994). This has been demonstrated for many human Ig isotypes, and the regulation by CD40 has been studied in some human B-cell lines. Control of isotype switching appears to be modulated, at least partly, through CD40-induced activation of the transcription factor nuclear factor (NF)-κB. The induction of NF-κB plays a crucial role in a number of CD40-mediated events, and the way in which CD40 clustering triggers the activation of NF-κB will be discussed below.

Another pivotal event in the life history of a B cell is the expression of Fas. It has been shown that the expression of CD154 on the activated helper T cell, in turn, induces Fas on the CD40 responsive B cell (Garrone 1995; Lagresle 1996). Since the T cell also expresses FasL, the B cell is killed unless membrane Ig (mIg) is cross-linked. Upon mIg engagement, antigen-reactive B cells are triggered to expand and differentiate by $CD154^+$ T cells. Therefore, while CD40 engagement alone can induce a wide spectrum of B-cell activities in vitro, B cells that are triggered in vivo by CD154 are those B cells that have mIg engaged by the cognate antigen. This model provides a compelling explanation for why B cells are not polyclonally activated by CD154-expressing helper T cells.

In addition to the role that B cells play in producing Ig, B cells play an important function in antigen presentation. As B cells are activated by CD40, they mature as antigen-presenting cells (APCs) by upregulating CD23, B7.1, B7.2, intercellular-adhesion molecule (ICAM)-1 and CD44, (see review by HATHCOCK 1996). B7.1 and B7.2 are likely the more important co-stimulatory molecules for APC functions, based on functional studies in genetically-deficient mice and blocking studies in vivo and in vitro (SHARPE 1995). While a great deal is known concerning the function of B7.1 and B7.2 in T-cell co-stimulation, the molecular basis for the regulation of B7.1 and B7.2 by CD40 is largely unresolved. Regulation by CD40 of B7.1 and B7.2 is paramount in eliciting immune response because it has been shown that blockade of CD40-induced B7 upregulation can ablate (or even tolerize) immune responses (BUHLMANN 1995; PARKER 1995).

## 1.3 CD40 and Its Ligand in the Regulation of Cell-Mediated Immunity

### 1.3.1 A Role in the Regulation of Antigen Presentation

In addition to the role of CD40 on B cells, CD40 also plays an important role in the regulation of macrophage and dendritic cell (DC) APC function. It has been shown that engagement of CD40 on macrophages can enhance APC activity and elicit a spectrum of pro-inflammatory and inflammatory cytokines (KIENER 1995). Furthermore, activation of DCs via CD40 can enhance their APC activity. Studies have shown that CD40 activation of DCs can induce the production of IL12, an important cytokine that can trigger a cascade of responses leading to heightened inflammatory (Th1-type) T-cell responses (BANCHEREAU 1995; CELLA 1996).

Early studies on the role of CD154 on B cells led to some important observations regarding the function of CD40 signaling in regulating cell-mediated immunity (CMI). Studies in our lab have shown that CD154 triggering of B cells is essential for B cells to become competent APCs (BUHLMANN 1995). We have shown that the engagement of TCR by peptide-major histocompatibility complex II induces the upregulation of CD154 in a CD80/CD86-independent manner (ROY 1995). CD154 on T cells triggers CD40 signaling on B cells, causing upregulation of CD80/86, and the heightened expression of CD80/86 by the cognate B cells facilitates the reciprocal triggering of the $CD4^+$ T cells via CD28. As such, the T cells are triggered to optimally expand and differentiate.

The importance of maintaining this sequential activation of the APCs and T cells during the creation of an immune response is demonstrated by the fact that blockade of CD154 does not result in "no response" but results in actual T-cell anergy. Thus, allograft reactions can be prevented with the administration of anti-CD154 antibody and allogeneic B cells, and it is this technology that has led to successful tolerance to alloantigens and allogeneic pancreatic-islet transplantation (PARKER 1995).

While B cells have always been considered "non-professional" APCs, DCs have long been thought to have innate (mature) APC capacities. Studies from our lab, in a model of acute GVHD, questioned whether a host bearing non-professional and professional APCs could present alloantigen to adoptively transferred allogeneic T cells when CD154 was blocked. The observation that GVHD was totally inhibited (DURIE 1994) demonstrated that the antigen-presentation capacity of both non-professional and professional APCs was not innate and could be prevented by blocking CD40 signaling. Further studies on the role of CD40 on DCs has shown that expression of CD40 on DCs is essential for the generation of protective tumor immune responses to autologous tumors (MACKEY 1998).

### 1.3.2 CD40 and T-Cell Polarity (Th1 and Th2)

A common consequence of CD154 blockade is ablation of interferon (IFN)γ production by T cells (STUBER 1996; MACKEY 1998). This has been observed in autoimmunity models (STUBER 1996), models of infectious disease (MACATONIA 1995) and in studies focused on the development of protective immunity to tumors (MACKEY 1998). CD40- or CD154-deficient mice are unable to mount protective, inflammatory immune responses to Leishmania major (CAMPBELL 1996), but administration of recombinant IL12 improved immune protection. IL-12 is induced as a consequence of CD40-induced DC activation (CAUX 1994; SHU 1995) and can facilitate Th1 differentiation. Studies by STUBER et al. (1996) complement these findings in that administration of αCD154 mAb was found to block hapten-induced colitis, an experimental model for inflammatory bowel disease. Specifically, αCD154 blocked IFNγ production and increased IL-4 secretion by lamina propria $CD4^+$ T cells in this system, and recombinant IL-12 administration led to severe disease even in the presence of therapeutic αCD154. Finally, studies from our lab have shown that IL12 production by tumors can partially overcome the need for CD154–CD40 interactions in eliciting protective tumor immunity. Therefore, a consensus is building that CD40 is critical for the development of CMI, and one component of its regulation is through control of the expression of IL12 by DCs.

## 2 CD40 Signal Transduction

The cytoplasmic domain of CD40 is a relatively small 62 amino acid domain which contains no obvious signaling motifs and which, in the human, does not contain any tyrosine residues that might act as docking sites following tyrosine phosphorylation as has been shown for other cell-surface receptors. Without any intrinsic capacity to signal, CD40 and other members of this family must rely on the association with adaptor molecules to propagate their stimulus to the cytoplasm and nucleus. Early studies on CD40 signaling provided evidence that the threonine at amino-acid position 254 was crucial for some CD40-mediated events. Later studies (GOLDSTEIN 1996) have shown that threonine 254, as well as threonine 247 in the tail of CD40, may play important roles in the biological function of CD40. Data are emerging that suggest that there may be several important functional domains in the CD40 cytoplasmic domain that control distinct biological events and engage different signaling cascades. The data suggesting that these domains may be sites for binding different adaptor molecules that play a role in transducing CD40 signals are discussed below.

As discussed below, there have been numerous studies on early events following CD40 ligation, including our own (MARSHALL 1993; REN 1994; MORIO 1995); however, few of them can be linked by a common thread that would provide us with insights into the general pathway of CD40 signaling. More recent studies may have uncovered one pathway that is prominently triggered on CD154 binding.

## 2.1 TNF-Receptor-Associated Factors

The TNF-R family, although homologous within their extracellular domains, do not contain obvious signaling motifs in their cytoplasmic domains. Some of the family members (Fas, TNF-R1) have what has been described as a "death domain". This 80-amino-acid domain appears to be involved in the caspase-mediated induction of apoptosis following engagement of these receptors by their cognate ligands. This function, however, is restricted to a few members of the family. Given their importance in a wide variety of biological responses, the nature of the signaling mechanism of the non-death receptors, including CD40, has been the focus of a number of studies. The first TRAF-family members were identified through their interaction with the cytoplasmic domain of the 75-kDa TNF-R (TNF-R2) (ROTHE 1994). Structurally, the TRAF molecules can be divided into subdomains that have different functions. All members of the family are homologous within the C-terminal domain, which has been termed the TRAF domain. The TRAF domain is further subdivided into a C-terminal portion termed TRAF-C and a region called TRAF-N, which is closer to the amino terminus and which possesses a coiled-coil structure. The coiled-coil domain is believed to be involved with the interactions of the TRAF molecules with other proteins found in the TRAF-receptor complex (ROTHE 1995). The TRAF-C region is critical for the interaction of TRAF molecules with their respective cell-surface receptors and is required for the homotypic aggregation of the cytoplasmic TRAF molecules (CHENG 1995). All TRAF molecules except for TRAF1 contain an N-terminal RING finger and all possess several zinc fingers, although the number varies between family members (CAO 1996). These structural motifs have been identified as regions of protein–protein or protein–DNA interactions. Deletion of the RING and some of the zinc fingers can eliminate the ability of these molecules to promote downstream signals, (CHENG 1995; ROTHE 1995; NAKANO 1996). To date, six TRAF family members have been identified (Fig. 2).

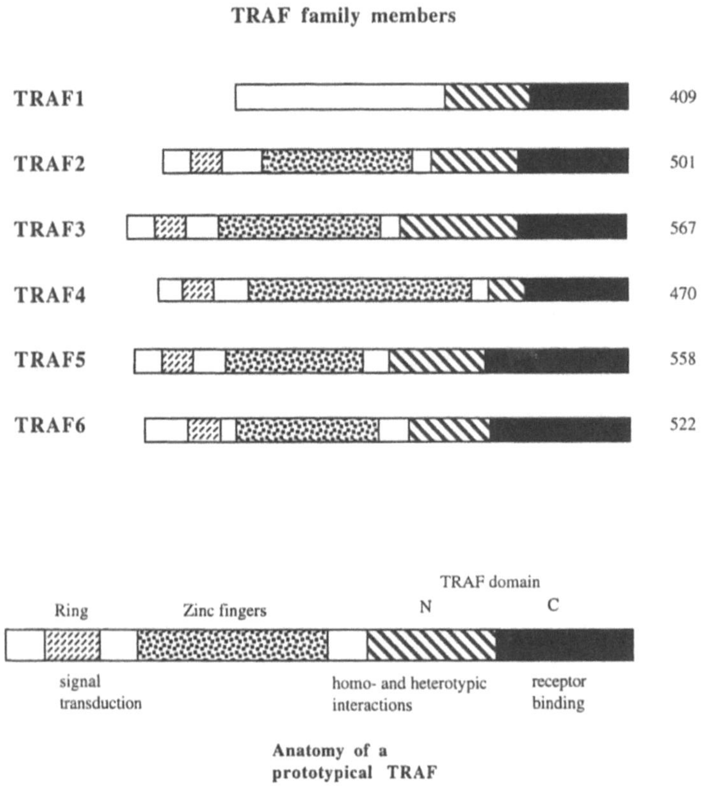


**Fig. 2.** Anatomy of TRAF proteins. The TRAF family of proteins bind to the cytoplasmic domain of CD40 and transduces signals to the intracellular machinery capable of mediating effector functions

### 2.1.1 TNF-Receptor-Associated Factor 1

TRAF1 and its homologue TRAF2 were isolated biochemically by means of their interaction with the TNF-R2 in a murine T-cell line CT6 (Rothe 1994). Although homologous within their TRAF domain, TRAF1, unlike TRAF2, does not contain an N-terminal RING finger. Northern-blot analysis of normal mouse tissues found that TRAF1 is expressed most highly in lung, testis and spleen. TRAF1 also appears to be induced in human B cells following infection by the Epstein-Barr virus (EBV; Mosialos 1995). Studies on TRAF1 and TRAF2 demonstrated that TRAF2 interacted directly with TNF-R2, and this interaction allowed association of TRAF1 with TNF-R2 (Rothe 1994). The association of TRAF1 with TRAF2 is mediated through the C-terminal TRAF-N domain. While TRAF1 requires TRAF2 for its interaction with TNF-R2 and possibly other receptors which utilize TRAF2, it appears to interact directly with CD30 (Gedrich 1996). Our own studies have shown that CD154 binding to CD40 induces the association of TRAF1 with the CD40-receptor complex (unpublished data).

### 2.1.2 TNF-Receptor-Associated Factor 2

Unlike TRAF1, TRAF2 (55kDa) is found in virtually every tissue by Northern-blot analysis. In addition to association with TNF-R2, TRAF2 has been found complexed to CD40, CD30, OX40, CD27 and TNF-R1 (ROTHE 1995; GEDRICH 1996; HSU 1996). TRAF2 has been shown to associate with a number of other proteins found in the cytosol, including TRAF family member-associated NFκB activator (TANK)/TRAF-interacting protein (CHENG 1996; ROTHE 1996), A20 (SONG 1996) and two members of the inhibitors-of-apoptosis-protein (IAP) family termed cIAP-1 and cIAP-2 (SHU 1996). Analysis of the various domains responsible for these associations found that the full TRAF domain, including both the TRAF-N and TRAF-C regions, was required for association with TNF-R2, while the homotypic association required only the TRAF-N region. TRAF2 appears to be strongly linked to the activation of NF-κB by members of the TNF-R family. In HEK 293 cells, overexpression of CD40 was shown to induce NF-κB without addition of CD154, presumably due to spontaneous receptor oligomerization. Co-transfection of TRAF2 along with CD40 enhanced this NF-κB activation. The N-terminal RING finger appeared to be critical for this activity, as co-transfection of a TRAF2 molecule lacking the first 97 residues, which includes this RING finger, completely blocked the CD40-induced NF-κB activity (ROTHE 1995). Deletion of the zinc-finger domains also might inhibit the TRAF2-induced activation of NF-κB (TAKEUCHI 1996). The link between TRAF2 and NF-κB will be discussed in greater detail below.

### 2.1.3 TNF-Receptor-Associated Factor 3

TRAF3 (also termed CD40-receptor-associated factor 1, CD40bp, CD40-associated polypeptide 1 and LMP1-associated protein) (HU 1994; CHENG 1995; SATO 1995; MOSIALOS 1995) was identified using the yeast two-hybrid system by several groups and has been shown to be associated with the cytoplasmic domain of CD40. In addition, TRAF3 has been shown to associate with CD30 and the latent membrane protein (LMP)-1 cell-surface protein expressed in EBV-transformed cells. One group has reported TRAF3 in association with both of the TNF receptors as well as the lymphotoxin-β receptor (LTβ-R), but this has not been found by other groups. In the mouse, the TRAF3 mRNA, like TRAF2, is found broadly expressed in all tissues studied. Interestingly reverse transcription polymerase chain reaction analysis of four B-cell lines found TRAF3 to be expressed in only two (HU 1994). Overexpression of TRAF3 does not induce NF-κB activation and, in fact, can block the NF-κB induction found with the overexpression of TRAF2 (ROTHE 1995).

### 2.1.4 TNF-Receptor-Associated Factor 4

This member of the family was identified through its homology in the C-terminal TRAF domain (REGNIER 1995). TRAF4 mRNA is expressed in some primary breast carcinomas but not in normal human tissues, and immunohistochemical staining of a breast carcinoma found that TRAF4 was found in the nuclei of malignant cells. This nuclear staining is interesting in light of the finding that some RING-finger-containing proteins are transcriptional activators. TRAF4 has not been shown to be associated with any cell-surface receptor.

### 2.1.5 TNF-Receptor-Associated Factor 5

TRAF5 was identified by two separate groups, one using the yeast two-hybrid system with the mouse CD40 cytoplasmic domain as bait (Ishida 1996; Nakano 1996). Both groups found TRAF5 broadly expressed in mouse tissues, with the highest levels of expression in the lung and moderate amounts in the thymus, spleen and kidney. One group found TRAF5 only associated with the LTβ-R and not CD40, TNF-R1 or -R2, Fas or the NGFR. Subsequent studies have shown that TRAF5 can associate with the cytoplasmic domains of CD30, OX40, and CD27 (Aizawa 1997; Akiba 1998; Kawamata 1998).

### 2.1.6 TNF-Receptor-Associated Factor 6

This member of the TRAF family was initially pulled from the expressed sequence tag database through its homology to TRAF2 within the TRAF domain (Cao 1996). Like TRAFs 2, 3 and 5, the mRNA is found in a broad range of tissues. TRAF6 was initially shown to be involved in IL-1 receptor (IL-1R) signaling, most likely due to an interaction with the serine/threonine kinase IL-1 receptor-associated kinase (IRAK). A later study using the CD40 cytoplasmic domain in a yeast two-hybrid screen demonstrated that TRAF6 could associate with CD40 but at a site more membrane proximal than either the TRAF2 or TRAF3 binding sites (Ishida 1996). At this time, there appear to be at least two functional domains in the CD40 tail: the domain that binds TRAFs 2, 3 and 5 and perhaps another domain that binds TRAF6. Of course, the other fascinating aspect of TRAF6 is that it can bind a serine-threonine kinase, IRAK. It is not known whether IRAK is associated with CD40 or involved in downstream signaling following engagement of CD40.

## 2.2 CD154-Induced CD40-Receptor Assembly

Upon binding of CD154 to its receptor, a number of changes in receptor behavior occur. CD154, due to its multimeric nature, aggregates CD40 and induces clustering. It is believed that the clustering of CD40 results in a multimeric site for the recruitment of TRAFs from the cytosol. We have shown that CD154 will induce the recruitment of TRAFs 1, 2, 3 and 6 to CD40 in a human B-cell line (Kuhne 1997). Recruitment of the TRAFs to the receptor complex occurs within minutes after ligand binding to the receptor. The recruitment of TRAF2 and TRAF3 to the receptor complex is dependent on the P-X-Q-X-T motif of the CD40 cytoplasmic tail, while the binding of TRAF6 is more N-terminal to the TRAF2/3 binding site. Mutations in the P-X-Q-X-T motif which ablate TRAF2/3 binding also impair a number of CD40-induced biological events (Inui 1990). These data suggest that the recruitment of TRAF2 or 3 is critical in CD40-induced events.

Engagement of mIg or triggering through the IL4-R can influence the recruitment of TRAFs to CD40. Engagement of mIgM on B cells results in the down-regulation of TRAF2 and, to a lesser extent, TRAF3 from the cytosol (Kuhne 1997). As a result, when CD40 is also engaged, there is a marked reduction in TRAF2 recruited to the receptor complex. In contrast to mIg engagement, preincubation of the B cells with IL-4 results in higher levels of TRAF2 recruited to CD40 upon addition of ligand. The changes in the levels of TRAF2 recruited to CD40 correlate directly with changes in the levels of Fas expression induced by CD154; i.e., reduced levels of TRAF2, caused by cross-linking of mIg, result in lower levels of Fas expression, while IL-4 enhances with CD154-induced Fas expression. Thus, the extent to which TRAF2 is recruited to the CD40-receptor complex correlates with at least some of the biological signals transduced by CD154.

## 2.3 CD40-Induced Signal Transduction and the Role of TRAFs

A common approach used to gain an understanding of the role of TRAFs in CD40 and TNFR signal transduction has been to co-transfect CD40 or TNFR and TRAF DNA (commonly into HEK 293 cells) to evaluate whether the TRAF protein can facilitate the transduction of a particular biochemical signal from the CD40 or TNFR. This approach has provided a wealth of data on the potential signaling capacities of different TRAF molecules. However, it has not always been easy to reconcile data between this approach and functional studies of TRAF molecules in non-manipulated systems or in mice genetically deficient or transgenic (Tg) in the expression of particular TRAF proteins.

### 2.3.1 The Role of TRAF2 in CD40 Signaling

NF-κB induction by TNF or CD40 engagement is a hallmark of TNF-R and CD40 signaling (Berberich 1994; Francis 1995) and, thus, the roles of TRAF molecules in NF-κB activation have been vigorously pursued. To determine the role of individual TRAF-family members in TNFR-mediated NF-κB activation, full-length TRAFs 1, 2 or 3 was transfected into 293 cells, a human embryonic kidney cell line. Transfection of TRAF1 or TRAF3 had no effect on either basal or TNFR-induced NF-κB activation (Rothe 1995). Transfection of full-length TRAF2 into 293 cells caused a strong upregulation of NF-κB activation in untreated cells. A dominant negative (DN) form of TRAF2 (in which the N-terminal RING finger was deleted) was also tested. The DN TRAF2 did not induce NF-κB and also inhibited induction by full-length TRAF2. These results established that TRAF2 was responsible for induction of NF-κB activation through TNF-R2 (CD40), and the observation that loss of the N-terminal RING finger could act in a DN manner indicated that this portion of TRAF2 was the effector of NF-κB activation. Use of TRAF2 as the bait in the yeast two-hybrid system identified a serine/threonine kinase termed NIK (for NF-κB-inducing kinase; MALININ 1997). Further studies using NIK as bait led to the recently described IκB kinase (IKK; REGNIER 1997). The IKKs (two, α and β have been identified) appear to be the kinases involved in the phosphorylation of IκB, a step critical to NF-κB activation. The connection between these molecules will be described in detail below.

While the 293 systems suggested an essential role of TRAF2 in NF-κB activation, studies in Tg mice, genetically disrupted mice and in B-cell lines concluded that TRAF2 was not essential in NF-κB activation. Choi and colleagues produced a Tg mouse over-expressing a DN construct (C-terminus) of TRAF2 (LEE 1997). Although the B cells from these mice demonstrated an impaired NF-κB activation following CD40 signaling, the primary defect was a complete loss in c-Jun N-terminal kinase (JNK) activation. $TRAF2^{-/-}$ mice have also been generated and have shown similar results (YEH 1997). Because the $TRAF2^{-/-}$ mice die 10 days following birth, it was not possible to study the response of mature B cells. However, embryonic fibroblast (EF) lines were generated from the $TRAF2^{-/-}$ mice. When these EF lines were triggered with TNF, NF-κB activation was impaired but not eliminated in a fashion similar to the CD40 signals found in the TRAF2 DN mice (YEH 1997). Like the TRAF2 DN Tg mouse, the $TRAF2^{-/-}$ demonstrated a complete impairment in JNK activation. Finally, studies in B-cell lines transfected with mutant forms of human CD40 (that cannot bind TRAF2 or TRAF3) also found that triggering CD40, which had a mutation at the threonine within the P-X-Q-X-T, still induced NF-κB, but at a significantly lower level (HSING 1997). Interestingly, this mutant was still capable of producing soluble IgM at levels equivalent to the wild-type (WT) transfectant (HOSTAGER 1996) as well as inducing expression of the cell-surface marker B7.1 (GOLDSTEIN 1996). However, it has been demonstrated that this mutation can interfere with other important CD40 growth signals (INUI 1990). These data suggest that TRAF2 is an essential mediator of JNK activation, and it's function in NF-κB activation may be redundant and compensated for by other members of the TRAF family, such as TRAF5 and TRAF6.

### 2.3.2 The Role of TRAF3 in CD40 Signaling

The importance of TRAF3 in mediating CD40 signals was initially inferred by the fact that TRAF3 bound to the CD40 cytoplasmic tail (HU 1994; CHENG 1995; SATO 1995). The most important fact established in these early studies was that interaction of CD40 with TRAF3 was dependent on the threonine residue at position 254, the same site involved with TRAF2 binding and a site critical in biological signaling by CD40 (INUI 1990). It was also shown that overexpression of a DN form of TRAF3 could block the CD40-induced expression of CD23 in Ramos B cells (CHENG 1995). However, a clear interpretation of the DN studies is difficult since the upregulation of CD23 on $TRAF3^{-/-}$ B cells from a $TRAF3^{-/-}$ mouse was similar to that found in wild-type mice on CD40 engagement (XU 1996). The most likely conclusion of the DN study was that DN TRAF3 displaced TRAF2 from the receptor and blocked the upregulation of CD23.

Further studies of the TRAF3$^{-/-}$ B cells demonstrated that B cells isolated from these mice were able to proliferate in response to CD40 signaling. In vivo, the TRAF3$^{-/-}$ mice mounted normal humoral immune responses to T-independent antigens, but their response to T-dependent antigens was impaired. The authors suggested a possible defect within the T-cell compartment in the TRAF3$^{-/-}$ mice that led to a defect in TD immunity; however, they had no definite evidence for this conclusion. The work to date suggests that, while TRAF3 is associated with CD40, its role as a mediator of CD40 signals is unclear.

### 2.3.3 The Role of TRAF5 and TRAF6 in CD40 Signaling

While TRAF2 has been directly shown to be involved in some aspects of CD40 signaling, recent work has implicated other members of the TRAF family as possibly being involved in this signaling pathway as well. TRAF5 was capable of inducing NF-κB activation, and a DN construct could partially block NF-κB activation induced by overexpression of the LTβ-R in 293 cells. A second group, which isolated TRAF5 using the CD40 tail in a yeast two-hybrid screen, found that overexpression of a DN molecule in WEHI-231 cells could block CD23 upregulation, similar to the results observed with TRAF3 (Ishida 1996). It was not demonstrated that WEHI-231 cells express native TRAF5, so the DN effect seen could be due to displacement of another TRAF molecule such as TRAF2. Studies on the interaction of TRAF5 with members of the TNF-R family indicate that in addition to the P-X-Q-X-T site used by TRAF2 and TRAF3, TRAF5 can bind to another site within the cytoplasmic tails of both CD30 and CD40, which is rich in acidic amino acids. This alternative site might allow TRAF5 to mediate the activation of NF-κB via CD40 in the absence of TRAF2 recruitment (Gedrich 1996).

The most recently described TRAF-family member is TRAF6. TRAF6 transfection into 293 cells induced NF-κB in a fashion similar to those of TRAF2 and TRAF5 (Cao 1996; Ishida 1996). A DN portion of the molecule had no effect on TNF-induced NF-κB activity in 293 cells (Cao 1996). In a variant 293 line which overexpresses the type-I IL-1R, the TRAF6 DN could inhibit IL-1-induced NF-κB activity (Cao 1996). Although TRAF6 was not found complexed to the IL-1R, it was found in association with a serine/threonine kinase known as IRAK, which has been shown to be recruited to the IL-1R and to become autophosphorylated upon challenge with IL-1. The TRAF6/IRAK association was also IL-1 dependent (Cao 1996).

Shortly after the description of TRAF6 as a mediator of IL-1 signals, it was shown that TRAF6 was part of the CD40 signaling complex (Ishida 1996). Unlike TRAF2 or TRAF3, TRAF6 was still bound to the CD40 tail when the threonine at 254 was converted to alanine although to a lesser extent than with WT tail. This is a very important observation, since it may define a new TRAF-binding domain of CD40. Deletion of residues 247–277, which includes the proposed P-X-Q-X-T TRAF2/3-binding site, only moderately reduced TRAF6 binding, while deletion of residues 231–277 abolished it (Ishida 1996). Also, transfection of the truncated CD40 molecule in which residues 247–277 were eliminated still resulted in some NF-κB activation in response to CD154. This induction was not blocked by co-transfection of a DN TRAF2 construct but was inhibited by a TRAF6 DN (Ishida 1996). Thus, it appears that TRAF6 may also be part of the CD40-signaling complex although at a site more membrane proximal than the previously described TRAF-binding site. It has not been shown that IRAK is part of the CD40/TRAF6 complex. A recent study has found that TRAF6 could associate with NIK, the purported activator of the IKK pathway leading to NF-κB activation (see below). This study also found that, along with TRAF6, co-transfection of a NIK DN construct could block the TRAF6-induced upregulation of NF-κB in HEK 293 cells (Song 1997).

### 2.3.4 Other Associated Proteins

The CD40 receptor appears to be capable of interacting with several members (TRAFs 2, 3, 5 and 6) of the TRAF family, some of which appear to be able to induce NF-κB activation. It is not clear at this point whether binding of TRAF2 results in binding of TRAF1 in a manner similar to TNF-R2. The mechanism for transmitting a signal from the CD40/TRAF complex to the mediators of NF-κB activation remains unclear. Work with TRAF2 in the TNF-R2 system has identified some other proteins which could play a role in this signal cascade. Goeddel's group, which originally identified TRAF1 and TRAF2, also described two proteins within the TNF-R2/TRAF complex (Rothe 1995). These two proteins, termed cIAP1 and cIAP2 (for cellular IAPs), were found to be homologous to each other and to several proteins found within the baculovirus genome. In the viral genome, IAPs appear to serve to prevent premature cell death of the host, allowing for longer viral replication and more successful viral propagation. The cIAPs also appear to be part of the TNF-R1 complex, and association with either receptor requires the TRAF2-TRAF1 heterocomplex (Rothe 1995; Shu 1996). Although no function has been described for the cIAPs in the TNF-R2-signaling pathway and no corresponding proteins for the CD40 receptor have yet been described, this linkage of apoptosis antagonists with a TNF-R family member raises questions about a possible role for them in the well-described CD40-mediated rescue of GC B cells (Liu 1991). A recent study has identified a serine/threonine kinase, receptor-interacting protein 2 (RIP2), which is found in association with CD40 (McCarthy 1998). This interaction is mediated through the binding of RIP2 to TRAFs 1, 5 and 6. RIP2 was found to both activate NF-κB and induce apoptosis when transfected into HEK 293 cells (McCarthy 1998). RIP2's role in these functions in B cells remains to be described.

There are two other proteins that may play an accessory role in the CD40 pathway due to their interactions with various TRAF members. These are TNF-R1-associated death-domain protein (TRADD) and IRAK. TRADD, which associates with the "death domain" of TNF-R1 was also found to interact with the TRAF-C domain of TRAF2 (Hsu 1996). TRADD has been shown to be important for TNF-induced NF-κB activation via TNF-R1. Overexpression of the TRAF2 DN could block this activation, thus implicating TRAF2 as a part of both the TNF-R1 and R2 NF-κB pathways. TRADD was required to recruit TRAF2 to TNF-R1, and so could play a role in CD40-receptor assembly even though CD40 has no defined "death domain". The demonstration that TRAF6 could interact with both CD40 and IRAK, the kinase associated with IL-1 signals, could be very significant (Ishida 1996).

## 2.4 The Linkage Between the CD40-TRAF-Receptor Complex and NF-κB Activation

The unique role that CD40 plays in the regulation of immune responses has prompted studies to understand the signaling pathways that are elicited from this receptor. These have unequivocally established NF-κB to be a key regulator of CD40-induced B-cell activity (Berberich 1994; Francis 1995; Sarma 1995; Rothe 1995). NF-κB is a family of transcription factors regulated by its interaction with the inhibitor IκB (Baldwin 1996; Ghosh 1998; May 1998; Sha 1998). It has long been known that various stimuli, including ultraviolet radiation, viruses and inflammatory cytokines, could induce the degradation of IκB, thereby freeing NF-κB to translocate to the nucleus. As a result, NF-κB could bind specific κB sequences upstream of a variety of genes and induce their expression. Importantly, a large number of immune-relevant genes have been shown to be under the influence of NF-κB, including many cytokines.

NF-κB plays a central role in B-cell growth and differentiation. In response to CD154, binding of specific NF-κB subunits to the ε and γ1 promoters have been shown to be essential for Ig-germline transcription (Iciek 1997; Lin 1998). Recently, γ1-germline transcripts have been shown to be dependent on competition between different NF-κB subunits. Since the composition of the active dimer which binds DNA seems to be the major element in determining the outcome of NF-κB activation, targeted gene disruption of family members has been used to shed light on their distinct functions. Of these, reports on $p50^{-/-}$, Rel $A^{-/-}$ and c-$Rel^{-/-}$ mice show defects in isotype switching (Kontgen 1995; Sha 1995; Snapper 1996; Doi 1997). The $p50^{-/-}$ mouse exhibited decreased levels of IgA, IgE and IgG1. Reduced levels of IgG1 and IgA were observed with $RelA^{-/-}$ B cells, while IgG1 and IgG2a levels were reduced in c-$Rel^{-/-}$ mice. In contrast, B cells from $RelB^{-/-}$ mice showed defects in CD40-stimulated proliferation but normal capacity of Ig to undergo isotype switching (Snapper 1996). In addition, $p52^{-/-}$ mice fail to form GCs (Franzoso 1998; Caamano 1998), a phenotype also seen in mice deficient in CD40 or CD40 ligand (Kawabe 1994; Xu 1994). Collectively, these reports illustrate the requirement for variable NF-κB subunits in CD40-dependent B-cell activation and, further, reveal the inability of NF-κB-family members to compensate for deficient subunits.

In the past year, a number of labs have advanced the understanding of NF-κB regulation considerably by identifying a complex of proteins capable of signaling the degradation of IκB (DiDonato 1997; Mercurio 1997; Regnier 1997; Woronicz 1997; Zandi 1997). This 700- to 900-kDa-molecular-weight complex, named the IKK complex, has been shown to consist of two kinases that can phosphorylate S32 and S36 of IκB-α (Lee 1998), residues which, upon phosphorylation, had been previously demonstrated to signal the ubiquitination and subsequent degradation of IκB-α. IKK-α, initially cloned as conserved helix-loop-helix ubiquitous kinase, was the first kinase identified out of this complex. Structurally, it consists of an N-terminal kinase domain, a middle leucine zipper region and a C-terminal helix-loop-helix domain. It also contains a SXXXS kinase-activation loop motif, characteristic of proteins of the mitogen-activated-protein-2 kinase (MAP2K) family. Another kinase closely related to IKK-α was isolated from the complex and named IKK-β (Mercurio 1997; Woronicz 1997; Zandi 1997). IKK-α and IKK-β are thought to form homo- and heterotypic complexes and it has been suggested that stoichiometric differences may play a role in varied activity of the IKK complex. Just recently, a third protein, named IKK-γ or NF-κB essential modulator, has been revealed, but its function has not yet been elucidated (Rothwarf 1998; Yamaoka 1998). The in vivo function of each individual protein remains to be defined. Nonetheless, all of the IKK proteins seem to be necessary for cytokine- or CD154-induced NF-κB activation.

Other protein kinases have been associated with the IKK complex. One group that initially identified IKK-α did so by using a yeast two-hybrid screen for proteins which interact with the MAP3K homologue NIK (Regnier 1997). NIK was originally described as a TRAF2-interacting molecule whose overexpression activated NF-κB (Malinin 1997). Since NIK is a MAP3K homologue and both IKK-α and IKK-β exhibit similarity to the MAP2K proteins, it was proposed that NIK might be the kinase upstream of the IKKs. Indeed, one report claims that NIK is capable of phosphorylating IKK-α on S176, critical for its activation (Ling 1998). Another group shows that overexpression of NIK results in the activation of both IKK-α and IKK-β. In addition, another MAP3K family member involved in the JNK pathway, mitogen-activated-protein kinase/ERK kinase kinase (MEKK)-1, was found in the IKK complex (Mercurio 1997). Overexpression of MEKK-1 has been shown to lead to phosphorylation of IκB-α at S32 and S36 (Lee 1997, 1998; Nakano 1998). In contrast to NIK however, MEKK-1 reportedly activates IKK-β preferentially over IKK-α. Since so many diverse stimuli activate NF-κB, it is possible that different signals may lead to differences in the components or state of the IKK complex, effecting differential activation of NF-κB.

In addition to kinases, NF-κB RelA and the substrates IκB-α and IκB-β have been found with the IKK-α and IKK-β proteins (Mercurio 1997; Regnier 1997). Notably, a recent article demonstrated that the IκB proteins are phosphorylated preferentially when they are associated with the NF-κB dimers (Zandi 1998). This presents a model for termination of the NF-κB response. NF-κB, once activated, regulates itself by inducing transcription of the IκB gene. Newly made unbound IκB travels from the cytoplasm to the nucleus to halt the activity of DNA-bound NF-κB. The finding that IKK preferentially phosphorylates IκBs bound to NF-κB allows free, newly synthesized IκB to escape degradation while in transit to the nucleus.

Recent evidence supports phosphorylation as the mechanism of IKK regulation. Our lab has demonstrated that, in response to CD40 signaling, IKK-α levels remain constant over time of stimulation, while its ability to phosphorylate IκB-α at S32 and S36 peaks at 5min and decreases rapidly until returning to near-basal levels at approximately 1h. At this time, an IKK-interacting phosphatase that counteracts IKK phosphorylation has yet to be conclusively identified. Interestingly, a protein which reacts with antibodies raised against the dual-specificity phosphatase MAP kinase phosphatase-1 (MKP-1) was present in the IKK complex (MERCURIO 1997). This protein, however, does not appear to be MKP-1 (judging from its molecular weight) and has not yet been identified. In addition, the phosphatase PP2A is apparently capable of dephosphorylating IKK-α in vitro, but this molecule has not been shown to be present in the IKK complex in vivo (DIDONATO 1997).

Due to the many diverse stimuli that activate NF-κB, studies are now underway to determine whether the IKK complex is uniquely vital in mediating NF-κB activation. Because most studies on IKK reported thus far have been in non-lymphoid cells, using the TNF and IL-1 system, many groups are investigating the role of the IKK complex in various NF-κB-inducing systems. Recently, two viral proteins, human T-cell leukemia virus (HTLV)-1 Tax and EBV LMP-1, both of which activate NF-κB, have been shown to employ the IKK complex (GELEZIUNAS 1998; SYLLA 1998; YIN 1998). We have studied the NF-κB pathway activated through CD40 in human B cells. Though CD40, TRAFs 2 and 3, NIK and MEKK-1 all failed to coassociate with the IKK complex, we have established that IKK-kinase activity is induced upon CD40 cross-linking. Therefore, intermediate proteins linking the CD40 receptor to the IKK-α and IKK-β kinases remain elusive. RIP2, a novel protein kinase shown to associate with TRAFs 1, 5 and 6, has also been shown to associate with the CD40 receptor in transient transfection studies (MCCARTHY 1998). Though its overexpression induces NF-κB, it is not known if RIP2 is an essential intermediate in the NF-κB-signaling cascade.

## 2.5 The Activation of Other Signaling Cascades by CD40

### 2.5.1 The Induction of JNK

Engagement of CD40 on the surface of B cells induces rapid phosphorylation and activation of the JNK in murine and human primary cells and transformed cell lines (SAKATA 1995; SUTHERLAND 1996; LI 1996; BERBERICH 1996). The pathway of enzyme activation leading from the activated CD40/TRAF family complex to the activation of stress-activated protein kinase (SAPK)/JNK is not yet fully resolved. SAPK/extracellular-signal-regulated kinase (ERK) kinase-1 [SEK1, also called JNK kinase 1 (JNKK1), mitogen-activated-protein kinase kinase 4] was the first kinase identified that directly phosphorylates and activates JNK (SANCHEZ 1994; DERIJARD 1995), though it also is capable of activating p38 kinase. Chimeric mice which are SEK1 deficient in their lymphocyte compartment have been developed, and while a subset of T-cell responses are impaired without SEK1, CD40-dependent responses including B-cell proliferation, Ig isotype switching, and GC formation are all unimpaired by the lack of SEK1 (NISHINA 1997). Therefore, SEK1 is not likely to be a candidate for upstream activator of JNK in CD40-dependent activation. JNKK2 was cloned 3 years later as a kinase that specifically activates JNK but not p38, but the physiological induction of JNKK2 was not reported (LU 1997). A third, TNF-responsive JNKK enzyme was cloned by two groups as MKK7 in mice (MORIGUCHI 1997) and MKK7 in humans (FOLTZ 1998). The human MKK7 protein was reportedly significantly induced by engagement of CD40, in contrast to SEK1, which was only mildly induced.

A MAP3K that is known to induce SEK1 is MEKK1 (MINDEN 1994; YAN 1994), and it is reported to be inducible by signaling through CD40 (SAKATA 1995). MEKK1 shares homology within its active site with another upstream signaling kinase NIK (MALININ 1997; LEE 1998). MEKK1 has also demonstrated the ability to induce NF-κB through activation of the IKK-enzyme complex (LEE 1998; NAKANO 1998; KARIN 1998) and is the physiological activator of NF-κB utilized by HTLV-1 through the viral protein Tax (GELEZIUNAS 1998; YIN 1998). Thus, MEKK1 represents a point of either redundancy or bifurcation within the CD40 signal cascade, since MEKK1 can induce both the IKK and JNK enzymes (MINDEN 1994; YAN 1994).

Upstream activators of MEKK1 which respond to CD40 have been suggested in a number of conflicting reports. One particularly attractive possibility, given the role of CD40 in GC formation, is that the GC Kinase (GCK) may connect a CD40/TRAF complex to the JNK signal cascade. GCK has recently been reported to bind to both TRAF2 and MEKK1 and it is proposed that GCK directly activates MEKK1 in this environment (YUASA 1998).

Interestingly, a very recent report describes a scaffolding protein called JNK-interacting protein (JIP-1), which selectively binds particular members of a JNK signal cascade (WHITMARSH 1998). JIP-1 specifically binds JNK at one site, the CD40-inducible kinase MKK7 at an adjacent site, and then members of the upstream mixed-lineage-kinase (MLK) family at a third site. To date, the MLK family has not been investigated with respect to CD40 signaling.

TRAF-family members have been implicated in the induction of JNK by CD40. Disruption of TRAF2 by targeted gene knockout or expression of DN mutants also impairs the ability of a CD40 signal to induce the activity of JNK (LEE 1997; YEH 1997). It seems most likely, according to currently available evidence, that CD40 induces JNK by recruiting TRAF2 to the receptor tail, along with a TRAF2-associating MAP4K like GCK. Phosphorylation of a downstream MAP3K like MEKK1 would lead into the JNK specific cascade, perhaps associated with the JIP-1 scaffolding protein, composed of MKK7 and JNK, and leading finally to phosphorylation of the transcription factor AP-1. Enzyme-substrate relationships have yet to be defined for GCK, MEKK1 and MKK7.

### 2.5.2 The Induction of MAP Kinase p38

Signaling through CD40 on the surface of a murine or human primary B cell or B-cell line induces the phosphorylation and activation of the MAPK p38

(Sutherland 1996; Salmon 1997; Grammer 1998). One MAPKK that can activate p38 is the poly-specific JNKK SEK-1 (Derijard 1995; Lin 1995). The MAPKK enzymes which specifically phosphorylate and activate p38 are MKK3 and MKK6 (Han 1996; Raingeaud 1996). Upstream activators of the p38 kinase pathway are not very well documented, but may include MEKK1, transforming-growth-factor-β-activated kinase 1, apoptosis-signal-regulating kinase 1 and MAP three kinase (MTK)-1 (Yan 1994; Derijard 1995; Yamaguchi 1995; Ichijo 1997).

Few reports have directly evaluated the role of specific TRAF-family proteins in the induction of p38 MAPK. TRAF2 reportedly activates p38 by way of its association with the serine/threonine kinase RIP (Yuasa 1998). Similarly, TRAF3 may be involved in signaling to p38 (Grammer 1998).

The physiological role of p38 activation in response to CD40 remains unclear. Using the specific p38 inhibitor SB203580, p38 was shown to downregulate cytoplasmic IgM expression in the Ramos B-cell line (Grammer 1998). However, SB203580 was unable to block the rescue of apoptosis by αCD40 (Salmon 1997). CD40-driven p38 activation has been shown to be vital to induction of ICAM-1 surface expression, as well as CD40-dependent proliferation in primary human tonsillar B cells (Craxton 1998). Surprisingly, p38 activation was not required for CD40 signaling for expression of Fas, DR3, or cIAP2. It is clear from these studies that some, but not all, CD40-induced biological responses are dependent on the p38-signaling pathway. Since a major function of p38 activity is the translational regulation of cytokine production, studies on more quickly and tightly regulated CD40 responses may be more informative.

### 2.5.3 The Induction of MAP Kinase ERK

The ERK kinase cascade is the most thoroughly studied of the MAPK cascades. ERK1 and 2 are induced by a broad variety of growth factors and nonreceptor kinases (for review see Robbins 1994). In Ramos cells, a human B-cell line, Sakata et al. showed that CD40 engagement resulted in a very rapid (30s) activation of JNK, but no corresponding activation of ERK (Sakata 1995). This was in striking contrast to surface IgM cross-linking, which elicited a pronounced and rapid ERK response in the same cells. However, in primary cultures of murine splenic B cells, Li et al. were able to detect activation of ERK in response to CD40 over a slow time course (approximately 10min of induction), while the ERK response to surface IgM cross-linking was detectable within 15s (Li 1996). Studies of IgM-induced apoptosis in the murine B-cell line WEHI-231 have also shown that CD40 does not induce phosphorylation and activation of ERK2, but rather can mildly inhibit ERK2 induction through surface IgM (Lee 1998). In contrast to WEHI-231 cells, primary splenic murine B cells upregulate ERK strongly, suggesting that the differentiation state of a B cell may influence the signal pathways invoked by CD40 (Purkerson 1998). Which differentiation states of a B cell may allow CD40 to invoke ERK signaling is as yet unknown.

## 3 Conclusions

Although the interactions of the TRAF molecules with CD40 provide an initial picture of the signaling pathway, a number of questions and new directions are presented. As described, engagement of the receptors by their corresponding multimeric ligand results in clustering of the cell-surface receptors. This aggregation results in the interaction of various TRAF proteins associated with the cytoplasmic domain of the receptor. Studies on the Fas receptor, another member of the TNF-R family, have shown that multimerization of the receptor is required for recruitment of apoptosis-transducing molecules (KISCHKEL 1995). Also, it is not known whether these proteins remain associated with the receptor to act as docking sites for other signaling proteins following engagement, or whether they disassociate from CD40 and move to other sites within the cell. For instance, TRAF2 and TRAF3 have RING fingers which in other proteins have been shown to mediate DNA binding and transcriptional activation, although no such function has been ascribed to these two TRAF-family members. Also, studies have found that overexpression of CD30 along with TRAF2 results in proteolytic degradation of the TRAF2 (DUCKETT 1997). This degradation may be a means of turning off signals from the receptor. Long-term stimulation through CD40 results in loss of TRAF molecules from the cell (unpublished results). Description of the TRAF molecules has provided an initial step linking CD40 to downstream signaling cascades, but the nature of the exact linkage to these cascades is yet to be resolved.

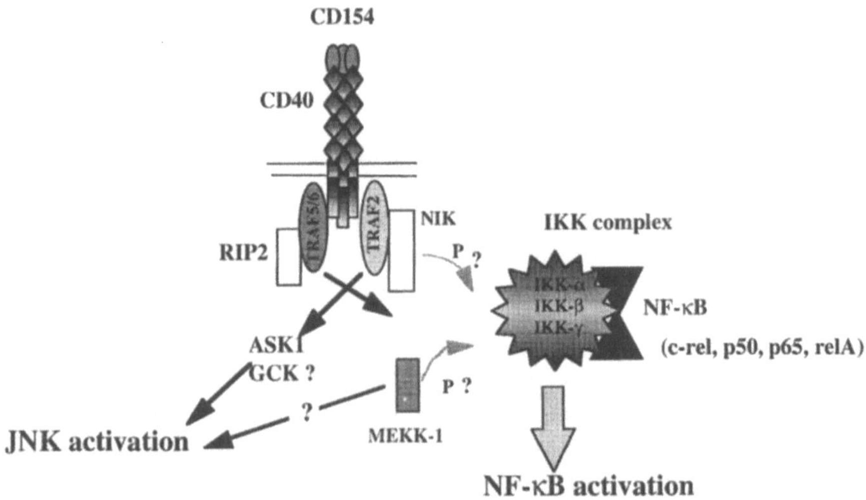


**Fig. 3.** The CD40 signaling cascade. The cytoplasmic signal from CD40 is transduced by kinase cascades. The assembly of TRAF proteins at the cytoplasmic tail of CD40 and the elicitation of nuclear signaling effects are well studied. However, the intermediate kinases which transduce these signals are not clearly established and are still under intense scrutiny

What is the linkage between TRAF2 and NF-κB? While NIK or MEKK1 have been proposed as the mediators of these signals, it has not been demonstrated that any of these signaling complexes are associated with the respective receptors. The activation of the JNK pathway through this receptor recruitment is similarly ill-defined. Overexpression of TRAF2 appears to result in a phenotype similar to that caused by triggering of the CD40 receptor, and this effect can be blocked by overexpression of TRAF3. Does the DN nature of TRAF3 in the 293 cells mean its function is to inhibit CD40-derived signals through TRAF2? Could the competing nature of TRAFs 2 and 3 be a means of controlling the cell's fate following CD40 engagement? An excess of TRAF2 might mean the balance is tipped towards the NF-κB pathway and perhaps isotype switching and antibody production, while a predominance of TRAF3 pushes the cell to proliferate or differentiate. The role of the other TRAF-family members remains to be elucidated, but given their ability to activate NF-κB and the residual ability of TRAF2 DN and –/– cells to trigger this pathway, it seems likely that they are involved in at least this activation pathway. Finally, in addition to the function of these molecules in facilitating the biochemical signaling from CD40, the role of these signal-transducing elements in the regulation of B-cell biology is yet to be resolved (Fig. 3).

## References


Aizawa S, Nakano H, Ishida T, Horie R, Nagai M, Ito K, Yagita H, Okumura K, Inoue J, Watanabe T (1997) Tumor necrosis factor receptor-associated factor (TRAF) 5 and TRAF2 are involved in CD30-mediated NF-kB activation. J Biol Chem 272:2042–2045

Akiba H, Nakano H, Nishinaka S, Shindo M, Kobata T, Atsuta M, Morimoto C, Ware CF, Malinin NL, Wallach D, Yagita H, Okumura K (1998) CD27, a member of the tumor necrosis factor receptor superfamily, activates NF-kappaB and stress-activated protein kinase/c-Jun N-terminal kinase via TRAF2, TRAF5, and NF-kappaB-inducing kinase. J Biol Chem 273:13353–13358

Alderson MR, Armitage RJ, Tough TW, Strockbine L, Fanslow WC, Spriggs MK (1993) CD40 expression by monocytes: Regulation by cytokines and activation of monocytes by the ligand for CD40. J Exp Med 178:669–674

Allen RC, Armitage RJ, Conley ME, Rosenblatt H, Jenkins NA, Copeland NG, Bedell MA, Edelhoff S, Disteche J, Simoneaux DK, Fanslow WC, Belmont J, Spriggs MK (1993) CD40 ligand gene defects responsible for X-linked hyper-IgM Syndrome. Science 259:990–993

Aloisi F, Ria F, Penna G, Adorini L (1998) Microglia are more efficient than astrocytes in antigen processing and in Th1 but not Th2 cell activation. J Immunol 160:4671–4680

Armitage RJ, Fanslow WC, Strockbine L, Sato TA, Clifford KN, Macduff BM, Anderson DM, Gimpel SD, Davis-Smith T, Maliszewski CR, Clark EA, Smith CA, Grabstein KH, Cosman D, Spriggs MK (1992) Molecular and biological characterization of a murine ligand for CD40. Nature 357:80–82

Aruffo A, Farrington M, Hollenbaugh D, Li X, Milatovich A, Nonoyama S, Bajorath J, Grosmaire LS, Stenkamp R, Neubauer M, Roberts RL, Noelle RJ, Ledbetter JA, Francke U, Ochs HD (1993) The CD40 ligand, gp39, is defective in activated T cells from patients with X-linked hyper-IgM syndrome. Cell 72:291–300

Bajorath J, Marken JS, Chalupny NJ, Spoon TL, Siadak AW, Gordon M, Noelle RJ, Hollenbaugh D, Aruffo A (1995) Analysis of gp39/CD40 interactions using molecular models and site-directed mutagenesis. Biochemistry 34:9884–9892

Baldwin AS Jr (1996) The NF-kappa B and I kappa B proteins: new discoveries and insights. [Review] [180 refs]. Annu Rev Immunol 14:649–683

Banchereau J, Bazan F, Blanchard D, Briere F, Galizzi JP, van Kooten C, Liu YJ, Rousset F, Saeland S (1994) The CD40 antigen and its ligand. [Review]. Annu Rev Immunol 12:881–922
Banchereau J, Dubois B, Fayette J, Burdin N, Briere F, Miossec P, Rissoan MC, van KC, Caux C (1995) Functional CD40 antigen on B cells, dendritic cells and fibroblasts. Adv Exp Med Biol 378:79–83
Berberich I, Shu G, Siebelt F, Woodgett JR, Kyriakis JM, Clark EA (1996) Cross-linking CD40 on B cells preferentially induces stress-activated protein kinases rather than mitogen-activated protein kinases. EMBO J 15:92–101
Berberich I, Shu GL, Clark EA (1994) Cross-linking CD40 on B cells rapidly activates nuclear factor-kappa B. J Immunol 153:4357–4366
Buhlmann JE, Foy TM, Aruffo A, Crassi KM, Ledbetter JA, Green WR, Xu JC, Shultz LD, Roopesian D, Flavell RA, Fast L, Noelle RJ, Durie FH (1995) In the absence of a CD40 signal, B cells are tolerogenic. Immunity 2:645–653
Caamano JH, Rizzo CA, Durham SK, Barton DS, Raventos-Suarez C, Snapper CM, Bravo R (1998) Nuclear factor (NF)-kappa B2 (p100/p52) is required for normal splenic microarchitecture and B cell-mediated immune responses. J Exp Med 187:185–196
Campbell KA, Ovendale PJ, Kennedy MK, Fanslow WC, Reed SG, Maliszewski CR (1996) CD40 ligand is required for protective cell-mediated immunity to Leishmania major. Immunity 4:283–9
Cao Z, Xiong J, Takeuchi M, Kurama T, Goeddel DV (1996) TRAF6 is a signal transducer for interleukin-1. Nature 383:443–6
Caux C, Massacrier C, Vanbervliet B, Dubois B, Van KC, Durand I, Banchereau J (1994) Activation of human dendritic cells through CD40 cross-linking. J Exp Med 180:1263–1272
Cella M, Scheidegger D, Palmer-Lehmann K, Lane P, Lanzavecchia A, Alber G (1996) Ligation of CD40 on dendritic cells triggers production of high levels of interleukin-12 and enhances T cell stimulatory capacity: T-T help via APC activation. J Exp Med 184:747–752
Cheng G, Baltimore D (1996) TANK, a co-inducer with TRAF2 of TNF- and CD 40L-mediated NF-kappaB activation. Gene Dev 10:963–973
Cheng G, Cleary AM, Ye Z, Hong DI, Lederman S, Baltimore D (1995) Involvement of CRAF1, a relative of TRAF, in CD40 signaling. Science 267:1494–1498
Clark EA (1990) CD40: a cytokine receptor in search of a ligand. Tiss Antig 36:33–36
Craxton A, Shu G, Graves JD, Saklatvala J, Krebs EG, Clark EA (1998) p38 MAPK is required for CD40 induced gene expression and proliferation in B lymphocytes. J Immunol 161:3225–3236
Cronin DN, Stack R, Fitch FW (1995) IL-4-producing CD8+ T cell clones can provide B cell help. J Immunol 154:3118–3127
Derijard B, Raingeaud J, Barrett T, Wu IH, Han J, Ulevitch RJ, Davis RJ (1995) Independent human MAP-kinase signal transduction pathways defined by MEK and MKK isoforms [published erratum appears in Science (1995) 269:17]. Science 267:682–685
DiDonato JA, Hayakawa M, Rothwarf DM, Zandi E, Karin M (1997) A cytokine-responsive IkappaB kinase that activates the transcription factor NF-kappaB [see comments]. Nature 388:548–554
DiSanto JP, Bonnefoy JY, Gauchat JF, Fischer A, de Saint Basile G (1993) CD40 ligand mutations in x-linked immunodeficiency with hyper-IgM. Nature 361:541–543
Doi TS, Takahashi T, Taguchi O, Azuma T, Obata Y (1997) NF-kappa B RelA-deficient lymphocytes: normal development of T cells and B cells, impaired production of IgA and IgG1 and reduced proliferative responses. J Exp Med 185:953–961
Duckett CS, Thompson CB (1997) CD30-dependent degradation of TRAF2: implications for negative regulation of TRAF signaling and the control of cell survival. Gene Dev 11:2810–2821
Durie FH, Aruffo A, Ledbetter JA, Crassi KM, Green WR, Noelle RJ (1994) Antibody to the ligand of CD40, gp39, blocks the occurence of the acute and chronic forms of graft-versus-host disease. J Clin Invest 94:1333–1338
Foltz IN, Gerl RE, Wieler JS, Luckach M, Salmon RA, Schrader JW (1998) Human mitogen-activated protein kinase kinase 7 (MKK7) is a highly conserved c-Jun N-terminal kinase/stress-activated protein kinase (JNK/SAPK) activated by environmental stresses and physiological stimuli. J Biol Chem 273:9344–9351
Foy T, Aruffo A, Bajorath J, Buhlmann JE, Noelle RJ (1996) Immune regulation by CD40 and its ligand gp39. Ann Rev Immunol 14:591–617
Foy TM, Aruffo A, Ledbetter JA, Noelle RJ (1993) In vivo CD40-gp39 interactions are essential for thymus-dependent immunity. II Prolonged in vivo suppression of primary and secondary humoral immune responses by an antibody targeted to the CD40 ligand, gp39. J Exp Med 178:1567–1575
Foy TM, Laman JD, Ledbetter JA, Aruffo A, Claassen E, Noelle RJ (1994) gp39-CD40 interactions are essential for germinal center formation and the development of B cell memory. J Exp Med 180: 157–164

Francis DA, Karras JG, Ke XY, Sen R, Rothstein TL (1995) Induction of the transcription factors NF-kappa B, AP-1 and NF-AT during B cell stimulation through the CD40 receptor. Int Immunol 7: 151–161
Franzoso G, Carlson L, Poljak L, Shores EW, Epstein S, Leonardi A, Grinberg A, Tran T, Scharton-Kersten T, Anver M, Love P, Brown K, Siebenlist U (1998) Mice deficient in nuclear factor (NF)-kappa B/p52 present with defects in humoral responses, germinal center reactions, and splenic microarchitecture. J Exp Med 187:147–159
Garrone P, Neidhardt EM, Garcia E, Galibert L, Van KC, Banchereau J (1995) Fas ligation induces apoptosis of CD40-activated human B lymphocytes. J Exp Med 182:1265–1273
Garside P, Inguilli E, Merica RR, Johnson JG, Noelle RJ, Jenkins MK (1998) Cognate interactions between antigen-specific CD4 T cells and B cells occur at the edge of the lymph node follicles and precede CD40L-dependent B cell clonal expansion. Science:in press
Gauchat JF, Henchoz S, Fattah D, Mazzei G, Aubry JP, Jomotte T, Dash L, Page K, Solari R, Aldebert D et al (1995) CD40 ligand is functionally expressed on human eosinophils. Eur J Immunol 25: 863–865
Gauchat JF, Henchoz S, Mazzei G, Aubry JP, Brunner T, Blasey H, Life P, Talabot D, Flores RL, Thompson J (1993) Induction of human IgE synthesis in B cells by mast cells and basophils. Nature 365:340–343
Gedrich RW, Gilfillan MC, Duckett CS, Van Dongen JL, Thompson CB (1996) CD30 contains two binding sites with different specificities for members of the tumor necrosis factor receptor-associated factor family of signal transducing proteins. J Biol Chem 271:12852–12858
Geleziunas R, Ferrell S, Lin X, Mu Y, Cunningham ET, Grant M, Connelly MA, Hambor JE, Marcu KB, Greene WC (1998) Human T-cell leukemia virus type 1 tax induction of nf-kappab involves activation of the IkappaB kinase alpha (IKKalpha) and IKKbeta cellular kinases. Mol Cell Biol 18:5157–5165
Ghosh S, May MJ, Kopp EB (1998) NF-kB and rel proteins: evolutionarily conserved mediators of immune responses. Annu Rev Immunol 16:225–260
Goldstein MD, Watts TH (1996) Identification of distinct domains in CD40 involved in B7–1 induction or growth inhibition. J Immunol 157:2837–2843
Grammer AC, Bergman MC, Miura Y, Fujita K, Davis LS, Lipsky PE (1995) The CD40 ligand expressed by human B cells costimulates B cell responses. J Immunol 154:4996–5010
Grammer AC, Swantek JL, McFarland RD, Miura Y, Geppert T, Lipsky PE (1998) TNF receptor-associated factor-3 signaling mediates activation of p38 and Jun N-terminal kinase, cytokine secretion, and Ig production following ligation of CD40 on human B cells. J Immunol 161:1183–1193
Gray D, Dullforce P, Jainandunsing S (1994) Memory B cell development but not germinal center formation is impaired by in vivo blockade of CD40-CD40-ligand interaction. J Exp Med:in press
Han J, Lee JD, Jiang Y, Li Z, Feng L, Ulevitch RJ (1996) Characterization of the structure and function of a novel MAP kinase kinase (MKK6). J Biol Chem 271:2886–2891
Hathcock KS, Hodes RJ (1996) Role of the CD28-B7 costimulatory pathways in T cell-dependent B cell responses. Adv Immunol 62:131–166
Hollenbaugh D, Mischel-Petty N, Edwards CP, Simon JC, Denfield RW, Kiener PA, Aruffo A (1995) Expression of functional CD40 by vascular endothelial cells. J Exp Med 182:33–40
Hostager BS, Hsing Y, Harms DE, Bishop GA (1996) Different CD40-mediated signaling events require distinct CD40 structural features. J Immunol 157:1047–1053
Hsing Y, Hostager BS, Bishop GA (1997) Characterization of CD40 signaling determinants regulating nuclear factor-kappa B activation in B lymphocytes. J Immunol 159:4898–4906
Hsu H, Shu HB, Pan MG, Goeddel DV (1996) TRADD-TRAF2 and TRADD-FADD interactions define two distinct TNF receptor 1 signal transduction pathways. Cell 84:299–308
Hu HM, O'Rourke K, Boguski MS, Dixit VM (1994) A novel RING finger protein interacts with the cytoplasmic domain of CD40. J Biol Chem 269:30069–30072
Ichijo H, Nishido E, Irie K, ten Dijke P, Saitoh M, Moriguchi T, Takagi M, Matsumoto K, Miyazono K, Gotoh Y (1997) Induction of apoptosis by ASK1, a mammalian MAPKKK that activates SAPK/JNK and p38 signaling pathways. Science 275:90–94
Iciek LA, Delphin SA, Stavnezer J (1997) CD40 cross-linking induces Ig epsilon germline transcripts in B cells via activation of NF-kappaB: synergy with IL-4 induction. J Immunol 158:4769–4779
Inui S, Kaisho T, Kikutani H, Stamenkovic I, Seed B, Clark EA, Kishimoto T (1990) Identification of the intracytoplasmic region essential for signal transduction through a B cell activation molecule, CD40. Eur J Immunol 20:1747–1753

Ishida T, Mizushima S, Azuma S, Kobayashi N, Tojo T, Suzuki K, Aizawa S, Watanabe T, Mosialos G, Kieff E, Yamamoto T, Inoue J (1996) Identification of TRAF6, a novel tumor necrosis factor receptor-associated factor protein that mediates signaling from an amino-terminal domain of the CD40 cytoplasmic region. J Biol Chem 271:28745–28748

Ishida T, Tojo T, Aoki T, Kobayashi N, Ohisho T, Watanbe T, Ymamoto T, Inoue J (1996) TRAF5, a novel tumor necrosis factor-receptor associated factor family protein meiadtes CD40 sginalling. Proc Natl Acad Sci USA 93:9437–9441

Jumper MD, Splawski JB, Lipsky PE, Meek K (1994) Ligation of CD40 induces sterile transcripts of multiple IgH chain isotypes in human B cells. J Immunol 152:438–445

Karin M, Delhase M (1998) JNK or IKK, AP-1 or NFkB, which are the targets for MEK kinase 1 action? Proc Natl Acad Sci USA 95:9067–9069

Karmann K, Hughes CW, Schechner J, Fanslow WC, Pober JS (1995) CD40 on human endothelial cells: Inducibility by cytokines and functional regulation of adhesion molecule expression. Proc Natl Acad Sci USA 92:4342–4346

Kawabe T, Naka T, Yoshida K, Tanaka T, Fujiwara H, Suematsu S, Yoshida N, Kishimoto T, Kikutani H (1994) The immune responses in CD40-deficient mice: impaired immunoglobulin class switching and germinal center formation. Immunity 1:167–178

Kawamata S, Hori T, Imura A, Takaori-Kondo A, Uchiyama T (1998) Activation of OX40 signal transduction pathways leads to tumor necrosis factor receptor-associated factor (TRAF) 2- and TRAF5-mediated NF-kappaB activation. J Biol Chem 273:5808–5814

Kiener PA, Moran DP, Rankin BM, Wahl AF, Aruffo A, Hollenbaugh D (1995) Stimulation of CD40 with purified soluble gp39 induces proinflammatory responses in human monocytes. J Immunol 155:4917–4925

Kischkel FC, Hellbardt S, Behrmann I, Germer M, Pawlita M, Krammer PH, Peter ME (1995) Cytotoxicity-dependent APO-1 (Fas/CD95)-associated proteins form a death-inducing signaling complex (DISC) with the receptor. EMBO J 14:5579–5588

Kontgen F, Grumont RJ, Strasser A, Metcalf D, Li R, Tarlinton D, Gerondakis S (1995) Mice lacking the c-rel proto-oncogene exhibit defects in lymphocyte proliferation, humoral immunity, and interleukin-2 expression. Gene Dev 9:1965–1977

Korthauer U, Graf D, Mages HW, Brieres F, Padayachee M, Malcolm S, Ugazio AG, Notarangelo LD, Levinsky RL, Kroczek A (1993) Defective Expression of T-cell CD40 ligand causes X-linked immunodeficiency with hyper-IgM. Nature 361:539

Kuhne MR, Robbins M, Hambor JE, Mackey MF, Kosaka Y, Nishimura T, Gigley JP, Noelle RJ, Calderhead DM (1997) Assembly and regulation of the CD40 receptor complex in human B cells. J Exp Med 186:337–342

Lagresle C, Mondiere P, Bella C, Krammer PH, Defrance T (1996) Concurrent engagement of CD40 and the antigen receptor protects naive and memory human B cells from APO-1/Fas-mediated apoptosis. J Exp Med 183:1377–1388

Lee FS, Hagler J, Chen ZJ, Maniatis T (1997) Activation of the IkappaB alpha kinase complex by MEKK1, a kinase of the JNK pathway. Cell 88:213–222

Lee FS, Peters RT, Dang LC, Maniatis T (1998) MEKK1 activates both IkB kinase alpha and IkB kinase beta. Proc Natl Acad Sci USA 95:9319–9324

Lee JR, Koretzky GA (1998) Extracellular signal-regulated kinase-2, but not cJun N-terminal kinase, activation correlates with surface IgM-mediated apoptosis in the WEHI 231 B cell line. J Immunol 161:1637–1644

Lee SY, Reichlin A, Santana A, Sokol KA, Nussenzweig MC, Choi Y (1997) TRAF2 is essential for JNK but not NF-kappaB activation and regulates lymphocyte proliferation and survival. Immunity 7: 703–713

Li YY, Baccam M, Waters SB, Pessin JE, Bishop GA, Koretzky GA (1996) CD40 ligation results in protein kinase C-independent activation of ERK and JNK in resting murine splenic B cells. J Immunol 157:1440–1447

Lin A, Minden A, Martinetto H, Claret FX, Lange-Carter C, Mercurio F, Johnson GL, Karin M (1995) Identification of a dual specificity kinase that activates the Jun kinases and p38-Mpk2. Science 268:286–290

Lin S, Wortis HH, Stavnezer J (1998) The ability of CD40L, but not lipopolysaccharide, to initiate immunoglobulin switching to immunoglobulin G1 is explained by differential induction of NF-kB/Rel proteins. Mol Cell Biol 18:5523–5532

Ling L, Cao Z, Goeddel DV (1998) NF-kappaB-inducing kinase activates IKK-alpha by phosphorylation of Ser-176. Proc Natl Acad Sci USA 95:3792–3797

Liu YJ, Mason DY, Johnson GD, Abbot S, Gregory CD, Hardie DL, Gordon J, MacLennan IC (1991) Germinal center cells express bcl-2 protein after activation by signals which prevent their entry into apoptosis. Eur J Immunol 21:1905–1910

Lu X, Nemoto S, Lin A (1997) Identification of c-Jun NH2-terminal protein kinase (JNK)-activating kinase 2 as an activator of JNK but not p38. J Biol Chem 272:24751–24754

Macatonia SE, Hosken NA, Litton M, Vieira P, Hsieh CS, Culpepper JA, Wysocka M, Trinchieri G, Murphy KM, O'Garra A (1995) Dendritic cells produce IL-12 and direct the development of Th1 cells from naive CD4+ T cells. J Immunol 154:5071–5079

Mackey MF, Gunn JRC, Maliszewski C, Kikutani H, Noelle RJ, Barth R (1998) Dendritic cells require maturation via CD40 to generate protective anti-tumor immunity. J Immunol 161:2094–2098

Malinin NL, Boldin MP, Kovalenko AV, Wallach D (1997) MAP3K-related kinase involved in NF-kappaB induction by TNF, CD95 and IL-1. Nature 385:540–544

Marshall LS, Aruffo A, Ledbetter JA, Noelle RJ (1993) The molecular basis for T cell help in humoral immunity: CD40 and its ligand, gp39. J Clin Immunol 13:165–174

May MJ, Ghosh S (1998) Signal transduction through NF-kappa B. [Review] [77 refs]. Immunol Today 19:80–88

McCarthy JV, Ni J, Dixit VM (1998) RIP2 is a novel NF-kB-activating and cell death-inducing kinase. J Biol Chem 273:16968–16975

Mercurio F, Zhu H, Murray BW, Shevchenko A, Bennett BL, Li J, Young DB, Barbosa M, Mann M, Manning A, Rao A (1997) IKK-1 and IKK-2: cytokine-activated IkappaB kinases essential for NF-kappaB activation [see comments]. Science 278:860–866

Minden A, Lin A, McMahon M, Lange-Carter C, Derijard B, Davis RJ, Johnson GL, Karin M (1994) Differential activation of ERK and JNK mitogen-activated protein kinases by Raf-1 and MEKK. Science 266:1719–1723

Moriguchi T, Toyoshima F, Masuyama N, Hanafusa H, Gotoh Y, Nishida E (1997) A novel SAPK/JNK kinase, MKK7, stimulated by TNFalpha and cellular stresses. EMBO J 16:7045–7053

Morio T, Hanissian S, Geha RS (1995) Characterization of a 23-kDa protein associated with CD40. Proc Natl Acad Sci USA 92:11633–11636

Mosialos G, Birkenbach M, Yalamanchili R, VanArsdale T, Ware C, Kieff E (1995) The Epstein-Barr virus transforming protein LMP1 engages signaling proteins for the tumor necrosis factor receptor family. Cell 80:389–399

Nakano H, Oshima H, Chung W, Williams-Abbott L, Ware CF, Yagita H, Okumura K (1996) TRAF5, an activator of NF-kappaB and putative signal transducer for the lymphotoxin-beta receptor. J Biol Chem 271:14661–14664

Nakano H, Shindo M, Sakon S, Nishinaka S, Mihara M, Yagita H, Okumura K (1998) Differential regulation of IkappaB kinase alpha and beta by two upstream kinases, NF-kappaB-inducing kinase and mitogen-activated protein kinase/ERK kinase kinase-1. Proc Natl Acad Sci USA 95:3537–3542

Nishina H, Bachmann M, Oliveira-dos-Santos AJ, Kozieradzki I, Fischer KD, Odermatt B, Wakeham A, Shahinian A, Takimoto H, Bernstein A, Mak TW, Woodgett JR, Ohashi PS, Penninger JM (1997) Impaired CD28-mediated interleukin 2 production and proliferation in stress kinase SAPK/ERK1 kinase (SEK1)/mitogen-activated protein kinase kinase 4 (MKK4)-deficient T lymphocytes. J Exp Med 186:941–953

Parker DC, Greiner DL, Phillips NE, Appel MC, Steele AW, Durie FH, Noelle RJ, Mordes JP, Rossini AA (1995) Survival of mouse pancreatic islet allografts in recipients treated with allogeneic small lymphocytes and antibody to CD40 ligand. Proc Natl Acad Sci USA 92:9560–9564

Purkerson JM, Parker DC (1998) Differential coupling of membrane Ig and CD40 to the extracellularly regulated kinase signaling pathway. J Immunol 160:2121–2129

Raingeaud J, Whitmarsh AJ, Barrett T, Derijard B, Davis RJ (1996) MKK3- and MKK6-regulated gene expression is mediated by the p38 mitogen-activated protein kinase signal transduction pathway. Mol Cell Biol 16:1247–1255

Regnier CH, Song HY, Gao X, Goeddel DV, Cao Z, Rothe M (1997) Identification and characterization of an IkappaB kinase. Cell 90:373–383

Regnier CH, Tomasetto C, Moog-Lutz C, Chenard MP, Wendling C, Basset P, Rio MC (1995) Presence of a new conserved domain in CART1, a novel member of the tumor necrosis factor receptor-associated protein family, which is expressed in breast carcinoma. J Biol Chem 270:25715–25721

Ren CL, Morio T, Fu SM, Geha RS (1994) Signal transduction via CD40 involves activation of lyn kinase and phosphatidylinositol-3-kinase, and phosphorylation of phospholipase Cg2. J Exp Med 179:673–680

Renshaw BR, Fanslow Wr, Armitage RJ, Campbell KA, Liggitt D, Wright B, Davison BL, Maliszewski CR (1994) Humoral immune responses in CD40 ligand-deficient mice. J Exp Med 180:1889–1900
Robbins DJ, Zhen E, Cheng M, Xu S, Ebert D, Cobb MH (1994) MAP kinases ERK1 and ERK2: pleiotropic enzymes in a ubiquitous signaling network. [Review] [145 refs]. Adv Cancer Res 63:93–116
Rothe M, Wong SC, Henzel WJ, Goeddel DV (1994) A novel family of putative signal transducers associated with the cytoplasmic domain of the 75 kDa tumor necrosis factor receptor. Cell 78: 681–692
Rothe M, Pan MG, Henzel WJ, Ayres TM, Goeddel DV (1995a) The TNFR2-TRAF signaling complex contains two novel proteins related to baculoviral inhibitor of apoptosis proteins. Cell 83:1243–1252
Rothe M, Sarma V, Dixit VM, Goeddel DV (1995b) TRAF2-mediated activation of NF-kappa B by TNF receptor 2 and CD40. Science 269:1424–1427
Rothe M, Xiong J, Shu HB, Williamson K, Goddard A, Goeddel DV (1996) I-TRAF is a novel TRAF-interacting protein that regulates TRAF-mediated signal transduction. Proc Natl Acad Sci USA 93:8241–8246
Rothwarf DM, Zandi E, Natoli G, Karin M (1998) IKK-gamma is an essential regulatory subunit of the IkappaB kinase complex. Nature 395:297–300
Roy M, Aruffo A, Ledbetter JA, Linsley P, Kehry M, Noelle RJ (1995) Studies on the independence of gp39 and B7 expression and function during antigen-specific immune responses. Eur J Immunol 25:596–603
Roy M, Waldschmidt T, Aruffo A, Ledbetter JA, Noelle RJ (1993) The regulation of the expression of gp39, the CD40 ligand, on normal and cloned CD4+ T cells. J Immunol 151:2497–2510
Sakata N, Patel HR, Terada N, Aruffo A, Johnson GL, Gelfand EW (1995) Selective activation of c-Jun kinase mitogen-activated protein kinase by CD40 on human B cells. J Biol Chem 270:30823–30828
Salmon RA, Foltz IN, Young PR, Schrader JW (1997) The p38 mitogen-activated protein kinase is activated by ligation of the T or B lymphocyte antigen receptors, Fas or CD40, but suppression of kinase activity does not inhibit apoptosis induced by antigen receptors. J Immunol 159:5309–5317
Sanchez I, Hughes RT, Mayer BJ, Yee K, Woodgett JR, Avruch J, Kyriakis JM, Zon LI (1994) Role of SAPK/ERK kinase-1 in the stress-activated pathway regulating transcription factor c-Jun. Nature 372:794–798
Sarma V, Lin Z, Clark L, Rust BM, Tewari M, Noelle RJ, Dixit VM (1995) Activation of the B-cell surface receptor CD40 induces A20, a novel zinc finger protein that inhibits apoptosis. J Biol Chem 270:12343–12346
Sato T, Irie S, Reed JC (1995) A novel member of the TRAF family of putative signal transducing proteins binds to the cytosolic domain of CD40. FEBS Lett 358:113–118
Sha WC (1998) Regulation of immune responses by NF-kappa B/Rel transcription factor [published erratum appears in J Exp Med (1998) 187:661]. [Review] [41 refs]. J Exp Med 187:143–146
Sha WC, Liou HC, Tuomanen EI, Baltimore D (1995) Targeted disruption of the p50 subunit of NF-kappa B leads to multifocal defects in immune responses. Cell 80:321–330
Sharpe AH (1995) Analysis of lymphocyte costimulation in vivo using transgenic and 'knockout' mice. Curr Opin Immunol 7:389–395
Shu HB, Takeuchi M, Goeddel DV (1996) The tumor necrosis factor 2 signal transducers TRAF2 and c-IAP1 are components of the tumor necrosis factor receptor 1 signalling complex. Proc Natl Acad Sci USA 93:13973–13978
Shu U, Kiniwa M, Wu CY, Maliszewski C, Vezzio N, Hakimi J, Gately M, Delepesse G (1995) Activated T cells induce interleukin 12 production by monocytes via CD40-CD40 ligand interaction. Eur J Immunol 25:1125–1128
Snapper CM, Rosas FR, Zelazowski P, Moorman MA, Kehry MR, Bravo R, Weih F (1996) B cells lacking RelB are defective in proliferative responses, but undergo normal B cell maturation to Ig secretion and Ig class switching. J Exp Med 184:1537–1541
Snapper CM, Zelazowski P, Rosas FR, Kehry MR, Tian M, Baltimore D, Sha WC (1996) B cells from p50/NF-kappa B knockout mice have selective defects in proliferation, differentiation, germ-line CH transcription, and Ig class switching. J Immunol 156:183–191
Song HY, Regnier CH, Kirschning CJ, Goeddel DV, Rothe M (1997) Tumor necrosis factor (TNF)-mediated kinase cascades: bifurcation of nuclear factor-kappaB and c-jun N-terminal kinase (JNK/SAPK) pathways at TNF receptor-associated factor 2. Proc Natl Acad Sci USA 94:9792–9796
Song HY, Rothe M, Goeddel DV (1996) The tumor necrosis factor-inducible zinc finger protein A20 interacts with TRAF1/TRAF2 and inhibits NF-kappaB activation. Proc Natl Acad Sci USA 93:6721–6725

Stamenkovic I, Clark EA, Seed B (1989) A B-lymphocyte activation molecule related to the nerve growth factor receptor and induced by cytokines in carcinomas. EMBO J 8:1403–1410

Stuber E, Strober W, Neurath M (1996) Blocking of CD40-CD40L interaction in vivo specifically prevents the priming of T helper 1 cells through the inhibition of interleukin 12 secretion. J Exp Med 183:693–698

Sutherland CL, Heath AW, Pelech SL, Young PR, Gold MR (1996) Differential activation of the ERK, JNK, and p38 mitogen-activated protein kinases by CD40 and the B cell antigen receptor. J Immunol 157:3381–3390

Sylla BS, Hung SC, Davidson DM, Hatzivassiliou E, Malinin NL, Wallach D, Gilmore TD, Kieff E, Mosialos G (1998) Epstein-Barr virus-transforming protein latent infection membrane protein 1 activates transcription factor NF-kB through a pathway that includes the NF-kB-inducing kinase and the IkB kinases IKKa and IKKb. Proc Natl Acad Sci USA 95:10106–10111

Takeuchi M, Rothe M, Goeddel DV (1996) Anatomy of TRAF2. Distinct domains for nuclear factor-kappaB activation and association with tumor necrosis factor signaling proteins. J Biol Chem 271:19935–19942

Tan L, Gordon KB, Mueller JP, Matis LA, Miller SD (1998) Presentation of proteolipid protein epitopes and B7–1-dependent activation of encephalitogenic T cells by IFN-gamma-activated SJL/J astrocytes. J Immunol 160:4271–4279

Van den Eertwegh AJM, Noelle RJ, Roy M, Shepherd DM, Aruffo A, Ledbetter JA, Boersma WJA, Claassen E (1993) In vivo CD40-gp39 interactions are essential for thymus-dependent immunity. I.CD40-gp39 interactions are essential for thymus dependent humoral immunity and identify sites of cognate interactions in vivo. J Exp Med 178:1555–1565

Whitmarsh AJ, Cavanagh J, Tournier C, Yasuda J, Davis RJ (1998) A mammalian scaffold complex that selectively mediates MAP kinase activation. Science 281:1671–1674

Woronicz JD, Gao X, Cao Z, Rothe M, Goeddel DV (1997) IkappaB kinase-beta: NF-kappaB activation and complex formation with IkappaB kinase-alpha and NIK [see comments]. Science 278:866–869

Xu J, Foy TM, Laman JD, Dunn JJ, Waldschmidt TJ, Elsemore J, Noelle RJ, Flavell RA (1994) Mice deficient for the CD40 ligand. Immunity 1:423–431

Xu Y, Cheng G, Baltimore D (1996) Targeted disruption of TRAF3 leads to postnatal lethality and defective T-dependent immune responses. Immunity 5:407–415

Yamaguchi K, Shirakabe K, Shibuya H, Irie K, Oishi I, Ueno N, Taniguchi T, Nishida E, Matsumoto K (1995) Identification of a member of the MAPKKK family as a potential mediator of TGF-beta signal transduction. Science 270:2008–2011

Yamaoka S, Courtois G, Bessia C, Whiteside ST, Weil R, Agou F, Kirk HE, Kay RJ, Israel A (1998) Complementation cloning of NEMO, a component of the IkappaB kinase complex essential for NF-kB activation. Cell 93:1231–1240

Yan M, Dai T, Deak JC, Kyriakis JM, Zon LI, Woodgett JR, Templeton DJ (1994) Activation of stress-activated protein kinase by MEKK1 phosphorylation of its activator SEK1. Nature 372:798–800

Yeh WC, Shahinian A, Speiser D, Kraunus J, Billia F, Wakeham A, de la Pompa JL, Ferrick D, Hum B, Iscove N, Ohashi P, Rothe M, Goeddel DV, Mak TW (1997) Early lethality, functional NF-kappaB activation, and increased sensitivity to TNF-induced cell death in TRAF2-deficient mice. Immunity 7:715–725

Yin MJ, Christerson LB, Yamamoto Y, Kwak YT, Xu S, Mercurio F, Barbosa M, Cobb MH, Gaynor RB (1998) HTLV-I Tax protein binds to MEKK1 to stimulate IkappaB kinase activity and NF-kappaB activation. Cell 93:875–884

Yuasa T, Ohno S, Kehrl JH, Kyriakis JM (1998) Tumor necrosis factor signaling to stress-activated protein kinase (SAPK)/Jun NH2-terminal kinase (JNK) and p38. J Biol Chem 273:22681–22692

Zandi E, Chen Y, Karin M (1998) Direct phosphorylation of IkappaB by IKKalpha and IKKbeta: discrimination between free and NF-kappaB-bound substrate. Science 281:1360–1363

Zandi E, Rothwarf DM, Delhase M, Hayakawa M, Karin M (1997) The IkappaB kinase complex (IKK) contains two kinase subunits, IKKalpha and IKKbeta, necessary for IkappaB phosphorylation and NF-kappaB activation. Cell 91:243–252

# B-Lymphocyte Signaling Receptors and the Control of Class-II Antigen Processing

N.M. Wagle[1], P. Cheng[2], J. Kim[2], T.W. Sproul[2], K.D. Kausch[2], and S.K. Pierce[2]



## 1 Introduction

The activation of B cells to proliferate and differentiate into clones of antibody-secreting plasma cells requires the binding of antigen to the B-cell antigen receptor (BCR). Cross-linking the BCR by the binding of bivalent or multivalent antigens initiates a signal-transduction cascade, which ultimately leads to the initiation of transcription of a variety of genes associated with B-cell activation (Pleiman et al. 1994). Although antigen-induced signaling through the BCR is necessary for B-cell activation, it is well appreciated that BCR signaling alone is not sufficient to induce proliferation and differentiation and that full B activation requires the interaction of B cells with antigen-specific helper T cells (Parker 1993). Although T cells express an antigen receptor, the T-cell antigen receptor (TCR), which is remarkably

[1] Lee Laboratories, 1475 Athens Highway, Grayson, GA 30017, USA
[2] Department of Biochemistry, Molecular Biology and Cell Biology, Northwestern University, Evanston, IL 60208, USA

similar in overall structure to the BCR, the recognition of foreign antigens by T cells and B cells is fundamentally different. The BCR engages antigen in its native conformation in a lock-and-key-like fit and consequently is exquisitely sensitive to small changes in the conformation of the antigen. In contrast, the TCR has no demonstrable ability to bind to antigens in their native structures but requires that the region of the antigen which contains the antigenic determinate be proteolytically cleaved from the native antigen and displayed bound to a major histocompatibility complex (MHC) class-II molecule (GERMAIN 1994). The foreign antigen is molecularly transformed into an antigenic-peptide class-II complex in an intracellular process termed antigen processing and presentation. B cells function as antigen-presenting cells (APCs) and, in fact, the interaction of B cells with helper T cells in the initiation of antibody responses is dependent on the ability of B cells to process and present antigens to helper T cells.

Given the central role antigen processing plays in initiating B-cell/T-cell interactions, it would seem likely that the assembly and cell-surface expression of antigenic-peptide class-II complexes is regulated in B cells, allowing B cells to control interactions with helper T cells. Indeed, as will be described in this review, evidence is accumulating that the antigen-processing function of B cells is regulated both developmentally and in response to an array of stimuli which serve to inform the B cells of a variety of factors, including the threat posed by the antigen, the B cell's progress in its interaction with helper T cells and the progression of the humoral immune response. Much of this information is received by B cells through cell-surface signaling molecules. To date, several signaling membrane proteins have been identified as influencing B-cell antigen processing, including the BCR itself, the Fc receptor (FcγRIIB1), the complement receptor CD21/CD19 complex, the class-II molecules and CD40.

Although, for the most part, the molecular mechanisms underlying the regulation of antigen processing in B cells have not been elucidated, the evidence thus far suggests that signals transduced through B-cell surface receptors modulate discrete steps in the BCR-mediated antigen-processing pathway. Indeed, recent results provide evidence that signals transduced by plasma-membrane receptors modulate both membrane movements within the antigen-processing pathway and the microenvironment of the endocytic compartments in which key antigen-processing events occur. Signaling receptors appear to affect antigen processing either directly, in which case the targets of the signal cascade are presumably components of the processing machinery, or indirectly primarily by modifying signaling through the BCR.

At present, a picture of B-cell antigen processing is emerging in which antigen presentation is a constitutive process, but one that is regulated by signals transduced through B-cell receptors to either promote assembly of peptide/MHC complexes or to turn off assembly when appropriate. Here, we will briefly review the current understanding of the BCR-mediated antigen processing pathway and then describe the present evidence that the BCR-mediated antigen-processing pathway is carefully regulated by a variety of factors.

## 2 The B-Cell Antigen Processing Pathway

In B cells, the processing of antigens is initiated by the binding of foreign antigens to the BCR (LANZAVECCHIA 1990; WATTS 1997). In this regard, B cells are as unique as APCs in their ability to discriminate foreign versus self antigens for subsequent processing. However, both macrophages and dendritic cells express receptors, including the macrophage Fc receptors and the dendritic cell mannose receptors, which have been shown to facilitate antigen processing, allowing a degree of selectivity in the antigen these cells will process (MANCA et al. 1991; JIANG et al. 1995; SALLUSTO et al. 1995). Following antigen binding, the BCR and bound antigen are internalized into early endosomes, and within minutes the antigen and the BCR enter a late endocytic compartment (WEST et al. 1994; SONG et al. 1995) where the antigen is degraded and the resulting peptides bind to newly synthesized class-II molecules. The fate of the BCR following antigen delivery has not yet been described.

The α and β chains of the class-II heterodimers are conventional membrane proteins, which are synthesized in the endoplasmic reticulum, where they rapidly associate with a trimer of another membrane protein termed the invariant chain (Ii; reviewed in CRESSWELL 1994). The Ii serves at least two functions in antigen processing: to direct the Ii class-II complexes to the endocytic system through information specified in its cytoplasmic tail (LOTTEAU et al. 1990; TEYTON et al. 1990) and to act as a surrogate peptide, filling the peptide-binding groove of the class-II molecules, stabilizing the class-II molecules and preventing peptide binding in the endoplasmic reticulum and Golgi (ROCHE and CRESSWELL 1990; ANDERSON and MILLER 1992). The majority of the newly synthesized class-II molecules appear to enter the endocytic peptide-loading compartments directly from the trans-Golgi network (PETERS et al. 1991; ODORIZZI et al. 1994; XU et al. 1995), although there is evidence that class-II molecules may be transported from the trans-Golgi network to the plasma membrane and are then retrieved to the peptide-loading compartments (KOCH et al. 1982; ROCHE et al. 1993). The Ii is proteolytically degraded once the Ii class-II complexes reach the peptide-loading compartment. Removal of the Ii occurs in a stepwise fashion, resulting in intermediate complexes of class-II molecules and fragments of Ii of decreasing molecular size [termed leupeptin-induced peptide (LIP) and small leupeptin-induced peptide (SLIP)] and a final intermediate composed of a class-II molecule and a small, 15- to 20-residue fragment of Ii [termed class-II-associated Ii peptide (CLIP) bound in the peptide-binding groove of the class-II molecule (CRESSWELL 1994). CLIP is replaced by an antigenic peptide in a process that is facilitated by a class-II-like molecule termed DM (ROCHE 1995). DM has been demonstrated in vitro to catalyze the release of CLIP from class-II molecules, allowing the binding of higher-affinity antigenic peptides (WEBER et al. 1996). The class-II αβ heterodimers are stabilized by the binding of antigenic peptides, and the stable dimers are transported, via a largely unknown pathway, to the plasma membrane for recognition by helper T cells.

It should be noted that antigen can also be taken up by B cells by fluid-phase pinocytosis; however, the processing and presentation of an antigen is thousands-fold more efficient when the antigen is internalized bound to the BCR (LANZAVECCHIA 1990; WATTS 1997). Thus, under physiological conditions in which antigen is likely to be limiting, it is unlikely that the fluid-phase route contributes significantly to B-cell antigen processing. In general, the fluid-phase pinocytosis pathway is less well characterized than the BCR-mediated pathway, and it is not known whether antigen taken up by fluid-phase pinocytosis traffics via the same intracellular route as antigen entering the cell bound to the BCR. Consequently, we will focus here primarily on the BCR-mediated antigen processing pathway and its regulation.

Recent progress has led to a fairly detailed description of the subcellular compartment in which peptide class-II complexes are assembled (XU and PIERCE 1995). This subcellular compartment was first described by immunoelectron microscopy by PETERS et al. (1991) as a multivesicular, late-endocytic, class-II-containing compartment which they termed the MIIC. Several laboratories, including our own, subsequently isolated and characterized the class-II peptide-loading compartments, here referred to as the IIPLC (QIU et al. 1994; RUDENSKY et al. 1994; TULP et al. 1994; WEST et al. 1994). Although there is some controversy over the nature of these compartments, there is a general agreement that these are late in the endocytic system, are enriched in lysosomal enzymes and lysosome-associated membrane proteins and do not contain significant amounts of the early endosome marker, the transferin receptor. The antigen-processing compartments appear to contain all the molecular machinery necessary to assemble peptide-class-II complexes including newly synthesized class-II molecules, proteases and DM (SANDERSON et al. 1994; SCHAFER et al. 1996). The peptide-loading compartment in B cells also contains another class-II-like molecule, DO, which recently was shown to be an inhibitor of DM function in vitro (LILJEDAHL et al. 1996; DENZIN et al. 1997). DO is of interest in considering the regulation of antigen processing in B cells because it is expressed exclusively in B cells and not in other APCs.

The characterization of the peptide-loading compartments raises several issues pertinent to the regulation of the assembly of peptide class-II complexes in B cells. The first is whether the IIPLC is a distinct endocytic compartment and, if so, whether the IIPLC is unique to B cells and/or to other class-II-expressing APCs. The present evidence indicates that the IIPLC may be a distinct compartment, but not unique to B cells or other APCs. The conclusion that the IIPLC is a distinct compartment is taken primarily from the fact that it has a distinctive multivesicular or multilaminar morphology (NEEFJES et al. 1990; PETERS et al. 1991, 1995) and, as will be discussed below, that the transport of the BCR and class-II molecules to the IIPLC is uniquely specified (MITCHELL et al. 1995; ALUVIHARE et al. 1997). The evidence that the IIPLC is not unique to B cells or other APCs comes from the observation that morphologically similar compartments form in non-APCs expressing transfected class-II molecules, and these compartments serve as sites for the assembly of functional peptide class-II complexes (CALAFAT et al. 1994). In addition, a recent, extensive, quantitative immunoelectron-microscopy study of the endocytic compartments of human and mouse B cells provides strong evidence that B cells use conventional endocytic compartments to assemble peptide class-II complexes rather than having developed unique antigen-processing compartments (Kleijmeer et al. 1997). Thus, the late-endocytic compartment in B cells, in which peptide class-II complexes assemble, likely serves similar functions for protein-complex assembly in other cell types. Because of the high level of activity of assembly of peptide class-II complexes in B cells, the IIPLC is an easily identifiable compartment in B cells, but nevertheless is not unique to B cells. Whether or not endocytic compartments of non-APC are identical to the IIPLCs in B cells and in other APCs is not known. It is certainly possible that B cells and other APCs modify endocytic compartments to endow them with functions that facilitate the specific task of assembly of peptide class-II complexes.

A second issue raised by the characterization of the peptide-loading compartments is whether the IIPLC is the only subcellular site in B cells in which peptide class-II complexes are assembled. Due to the plasticity of the endocytic system, it is difficult to precisely define individual compartments. However, using immunoelectron microscopy, Kleijmeir et al. (1997) subdivided the endocytic pathway in B cells into six subcompartments and, using an antibody specific for one particular abundant peptide class-II complex, found the majority of peptide class-II complexes in three of these, which were late in the endocytic system. Peptide class-II complexes were not found in early endosomes. In experiments in which the antigen was bound to the BCR and then chased into B cells through the early and late endosomes, we were unable to detect the formation of functional peptide class-II complexes capable of stimulating a T cell in any subcellular compartment other than in dense endocytic compartments (Qiu et al. 1994). Clearly, the late-endocytic compartments are important sites of peptide-class-II assembly. However, peptide class-II complexes have been demonstrated to form in early-endocytic structures (Castellino and Germain 1995; Escola et al. 1995; Griffin et al. 1997). In certain cases, processing in early endosomes involved class-II molecules cycling from the plasma membrane through early endosomes, was DM-independent and only occurred for a subset of epitopes within antigens which appear to be readily released by proteolysis. The degree to which class-II molecules bind peptides outside late-endocytic compartments and the question of whether assembly in all subcellular sites is similarly regulated remain to be determined.

A final issue with regard to the regulation of antigen processing is whether the transport pathways to the IIPLC from the trans-Golgi network or from the plasma membrane are unique. Specialized transport pathways would provide a means of controlling delivery of antigen and/or regulatory molecules to the IIPLC without affecting transport of housekeeping proteins or proteins destined for degradation to late endosomes or lysosomes. Concerning the transport from the trans-Golgi network to the IIPLC, current evidence indicates that Ii-class-II complexes are trafficked via the mannose-6-phosphate-receptor (MPR) pathway, but that DM is independently transported via a separate route. Using immunoelectron microscopy, Glickman et al. (1996) showed that class-II molecules exited the trans-Golgi network (TGN) in vesicles that were distinct from the clatherine-coated vesicles which transport the MPR to the endocytic system. Lui et al. (1998), using a dominant-negative mutant form of the hulls fragment of clathrin that disrupts clatherine-basket assembly, showed that DM trafficks via a clatherine-independent pathway to the IIPLC, whereas the transport of Ii class-II complexes is clatherine dependent, as is the transport of MPR. Consistent with these observations, we showed that the transport of DM, but not the Ii class-II complexes, was blocked by the PI-3 kinase inhibitor wortmannin, which prevents transport of the MPR from the TGN to the endocytic system (Song et al. 1997; Sproul, unpublished observations). These observations are of considerable interest, as they suggest that the movement of the catalyst DM, which facilitates peptide class-II assembly, to the IIPLC may be regulated independently of class-II transport. Concerning the transport of molecules from the plasma membrane to the IIPLC, as will be discussed below, transport of the BCR and bound antigen from the plasma membrane to the IIPLC appears to be uniquely specified and distinct, at least in part, from transport from the plasma membrane to lysosomes.

## 3 The Dual Function of the BCR in Antigen Processing

As described above, antigen processing in B cells is initiated by the binding of antigen to the BCR, which functions to transport bound antigen to the IIPLC. In addition, the BCR serves a second function in antigen processing, which is to signal for enhanced processing (Casten et al. 1985). As a consequence of the BCR's dual roles of antigen transport and signaling, BCR-mediated antigen processing is a highly efficient process. The antigen-transport function of the BCR was first suggested by the efficiency with which antigens that bound to the BCR were processed and presented compared with antigens taken up by fluid-phase pinocytosis (Lanzavecchia 1990; Watts 1997). Indeed, antigen-specific B cells require 1/1000 to 1/10,000 the amount of antigen for maximal presentation compared to B cells which are not specific for the antigen. Recently, the isolation and characterization of the subcellular compartments in which peptide class-II complexes are assembled, described above, made it possible to directly demonstrate that the BCR transports bound antigens from the plasma membrane to the processing compartments (West et al. 1994; Song et al. 1995). Tracking the movement of the BCR and bound antigen to the IIPLC also provided additional information concerning the content of the IIPLC. Recent studies from our laboratory, using horseradish-peroxidase-anti-immunoglobulin (Ig) as an antigen to catalyze chemical cross-linking of proteins present in the same subcellular compartment as the BCR, demonstrated that the antigen and the BCR traffic through transferrin-receptor-containing early endosomes and enter compartments that contain all the intermediates in the processing pathway, including newly synthesized class-II molecules, SLIP-, LIP- and CLIP class-II complexes and DM (Song et al. 1995; Cheng et al. 1999).

Immunoelectron microscopy supported this conclusion, showing rapid colocalization of the BCR and antigen in multi-vesicular, class-II-containing compartments.

In addition to the BCR's function in transporting antigen to the IIPLC, signaling through the BCR results in enhanced antigen processing. The effect of BCR signaling is manifested in several ways. Cross-linking the BCR stimulates the processing of antigens taken up by fluid-phase pinocytosis (CASTEN et al. 1985). Thus, B cells treated with anti-Ig to cross-link the BCR require 1/10 to 1/50 the amount of antigen to stimulate an antigen-specific T-cell hybrid than with untreated B cells. This effect of BCR cross-linking is dependent on B-cell kinase activity and is blocked by kinase inhibitors that block BCR signaling (XU et al. 1996; WAGLE et al. 1998). The stimulatory effect of BCR cross-linking on antigen processing is not dependent on B-cell proliferation and can be demonstrated in cells treated with DNA-synthesis inhibitors.

The observed effect of BCR signaling on the APC function of B cells is mirrored by biochemical changes in the IIPLC. The signal-transduction cascade initiated by cross-linking the BCR at the plasma membrane results in a rapid and transient change in specific subsets of regulatory proteins associated with the IIPLC, including phosphoproteins and low-molecular-weight guanosine triphosphatases (GTPases; XU et al. 1996). Within minutes following BCR cross-linking, the pattern of phosphoproteins associated with the IIPLC changes dramatically. Predominant phosphoproteins of approximately 30kDa, 45kDa and in the 50- to 60-kDa range were observed. The changes in phosphoprotein profiles associated with the IIPLC were blocked by kinase inhibitors that block BCR signaling, and were not accompanied by overall gross changes in the protein composition of the subcellular compartments. In addition, phosphoproteins associated with other subcellular compartments, including the early and late endosomes, Golgi and transport vesicles, were for the most part unchanged following BCR cross-linking.

Of considerable interest is the recent observation that a subset of the phosphoproteins that become associated with the IIPLC following BCR cross-linking become rapidly and transiently associated with the class-II molecules in the IIPLC, as evidenced by the ability to selectively co-immunoprecipitating these phosphoproteins along with class-II molecules (Kausch, unpublished observation). These phosphoproteins (43–45kDa and 50–60kDa) can be chemically cross-linked to the class-II molecules using the 3A° cross-linking reagent difluorodinitrobenzene and are distinct from the phosphorylated proteins associated with the BCR following BCR cross-linking. The dependence of the association of these phosphoproteins on BCR signaling was shown by the ability of kinase inhibitors that block BCR signaling to block the association of the phosphoproteins with the class-II molecules. Although the identity of these phosphoproteins is not yet known, we speculate that they may be part of a complex which assembles in response to BCR signaling in order to facilitate the loading of class-II molecules with peptides derived from the antigen bound to the BCR. Such a complex would be analogous to the 'TAP complex', which insures the efficient loading of class-I molecules with peptides in the endoplasmic reticulum (SADASIVAN et al. 1995). The TAP complex minimally contains the transporters associated with antigen processing (TAP), class-I heavy chains, β2m, the chaperone calreticulin and tapasin. At this point, we can only speculate as to the minimal components of a loading complex for the class-II molecules. Previous studies demonstrated that DM associates with class-II molecules and Ii class-II complexes in the IIPLC (Sanderson et al. 1996; Schafer et al. 1996). Such complexes may represent the core of a loading complex which functions in constitutive antigen processing in B cells. We envision that a peptide-loading complex in activated cells may also include the BCR and bound antigen and that this complex and its assembly may be key targets for regulation of antigen processing in B cells.

We also observed a selective change in the pattern of low-molecular-weight GTPases associated with the IIPLC following BCR cross-linking (Xu et al. 1996). These are of interest, as many of the proteins which mediate vesicular trafficking within the endocytic system are GTPases, such as the Rab proteins. The changes in the patterns of the low-molecular-weight GTPases are blocked by kinase inhibitors that block BCR signaling. The question of how the changes in the patterns of the low-molecular-weight GTPases relate to the effect of BCR signaling on antigen processing remains to be established. As will be described below, BCR-signaling alters the rate of trafficking of the BCR and antigen to the IIPLC. It is possible that the observed changes in low-molecular-weight GTPases associated with the IIPLC facilitate this accelerated trafficking. Similarly, as discussed above, DM and the Ii class-II complexes appear to traffic independently of class-II molecules to the IIPLC, offering the opportunity to regulate the movement of these molecules to the IIPLC. Conceivably, BCR signaling controls DM trafficking through alterations in the association of low-molecular-weight GTPases with the IIPLC.

Signals transduced by cross-linking the BCR also influence the internalization and trafficking of the BCR and bound antigen to the IIPLC. We first observed that conjugates of antigen covalently coupled to antibodies specific for Ig prepared with monovalent Fab anti-Ig were processed approximately 50- to 100-fold more efficiently than the same antigens taken up by fluid-phase pinocytosis (Casten and Pierce 1988). This observation indicated that cross-linking the BCR was not required for efficient antigen processing. However, antigen-anti-Ig conjugates prepared using bivalent $F(ab')_2$ anti-Ig were 10- to 50-fold more efficient than monovalent antigen-Fab-anti-Ig conjugates (Casten and Pierce 1988), indicating that BCR cross-linking augmented BCR-mediated processing. Investigating the phenomenon further, we determined that binding of bivalent antigens to the BCR resulted in a greater level and rate of internalization and degradation of the BCR and bound antigen than monovalent antigen binding (Song et al. 1995). Using subcellular fractionation, we showed that the BCR constitutively travels to the IIPLC in the absence of antigen or upon binding of monovalent antigen, but that cross-linking the BCR greatly accelerated the rate of transport without affecting the route of transport (Song et al. 1995). Using kinase inhibitors that block BCR signaling, we documented that the effect of BCR cross-linking on the degradation of the BCR and bound antigen and its trafficking to the IIPLC was dependent on kinases that participated in signaling cascades and was not due to the simple physical aggregation of the cross-linked BCR (Wagle et al. 1998). In contrast, the increase in the level and rate of BCR internalization from the plasma membrane appeared to be independent of BCR signaling. Consistent with these observations, Alavihara et al. (1997) recently showed that intracellular trafficking of antigen to the IIPLC by the BCR was accelerated compared with the trafficking of antigen by chimeric IgM molecules which contain targeting sequences from plasma-membrane proteins other than the BCR. The kinetics of BCR transport in the absence of signaling was not addressed by the authors; thus, it is not known whether the observed accelerated trafficking of the BCR required signaling. However, based on our findings, we would predict that the accelerated trafficking is dependent on BCR signaling.

Taken together, these results support a model for B-cell antigen processing in which resting B cells assemble peptide class-II complexes constitutively. The binding of bivalent or multivalent antigen to the BCR transduces signals which both accelerate the delivery of the BCR and bound antigen to the IIPLC and heighten antigen-processing activity in the IIPLC so that the B cells are able, in a concerted fashion, to efficiently process and present antigens bound to the BCR. In such a model, the BCR functions to communicate to the IIPLC, through a signal-transduction cascade, that a foreign antigen has been encountered and will be delivered momentarily for processing.

To function in its dual roles in antigen processing, the BCR must both specify the information to be targeted to the IIPLC and function in initiation of signal transduction. At present, the relationship between these functions and the structure of the BCR has not been fully delineated. In particular, the minimal BCR required for constitutive trafficking to the IIPLC versus accelerated, regulated trafficking is not known. The BCR complex is composed of the surface Ig (sIg) and a noncovalently associated disulfide-linked heterodimer of Igα and Igβ (reviewed in Reth 1992 and in this volume). The cytosolic domains of Igα and Igβ contain the immune-receptor-tyrosine activator motifs, or ITAMs, which become phosphorylated on BCR cross-linking and function to link the sIg to cytosolic tyrosine kinases in signal transduction (Pleiman et al. 1994). Mutational studies of the BCR provide evidence that the efficiency of antigen processing following antigen binding of the BCR can be influenced by the cytoplasmic domains of the Igα and Igβ proteins. Mutagenesis of the transmembrane regions of IgM which blocked association with Igα and Igβ resulted in loss of both signaling and endocytic function (Grupp et al. 1993). Both functions could be restored by the addition of the cytoplasmic tail of Igβ to the sIgM. Neuburger and Patel (1993) showed that the tyrosine residues in the ITAM sequences of the Igα and Igβ were not required for BCR-facilitated antigen processing, but whether the tyrosine residues were required for accelerated trafficking of the BCR was not tested.

Studies by Bonnerot et al. (1995) using chimeric molecules containing the extracellular domain of the FcγRII and Igβ or Igα cytoplasmic tail showed that Igα and Igβ were each sufficient to allow internalization, but chimeras containing Igβ were not targeted to processing compartments and cycled through the early endosomes, whereas Igα chimeras were correctly targeted. Cassard et al. (1998) recently provided evidence that only the Igα cytoplasmic domain contains a signal that allowed constitutive internalization of the BCR. ALAVIHARA et al. (1997), using a set of chimeric IgM molecules, showed that the Igα and Igβ cytoplasmic domains of the BCR were responsible for accelerated transport of antigen to the IIPLC. Other cytoplasmic tails, such as those derived from the low-density lipoprotein (LDL) or the class-I heavy-chain ultimately delivered antigen to the IIPLC but in a process which required considerable additional time, suggesting a different intracellular route. Moreover, the ability of LDL and the class-I heavy-chain chimeric IgM molecules, but not the wild-type IgM, to facilitate processing was dependent on the strength of the antigen binding to the IgM ectodomain. This observation suggests that the LDL and class-I heavy-chain cytoplasmic domains specified transport of antigen through a route distinct from that taken by the BCR, and one that requires additional time and high-affinity antigen binding. The membrane Ig (mIg) has also been implicated in specifying BCR-facilitated antigen processing. MITCHELL et al. (1995) found that a single point mutation in the transmembrane region of IgM abolished antigen processing, although association with Igα and Igβ was normal and the BCR was internalized and degraded. Thus, in the absence of correct targeting the BCR and bound antigen were presumably relegated to the lysosomal degradation pathway and the antigen was not transported to the IIPLC. The cytoplasmic domain of the mIgG heavy chain has also been implicated in specifying endocytosis and facilitating antigen processing independently of Igα and Igβ (KNIGHT et al. 1997).

Taken together, these results suggest a complex picture in which the domains of the BCR may not function independently, and more than one domain may specify targeting and function in signal transduction. It will be of considerable interest to determine which domains of the BCR specify correct targeting to the IIPLC and which are required for the accelerated trafficking following BCR cross-linking.

The observations of MITCHELL et al. (1995) and ALAVIHARA et al. (1997) are also significant in providing evidence that trafficking from the plasma membrane to the IIPLC is via a discrete pathway, distinct from that taken by plasma-membrane proteins destined for degradation in lysosomes. As discussed above, the existence of discrete pathways from the plasma membrane to the IIPLC provides supporting evidence that the IIPLC is a unique endocytic compartment, distinct from lysosomal degradation compartments.

## 4 Regulating BCR-Regulated Antigen Processing

As described above, the BCR, as a signaling receptor, has the ability to sense the presence of antigen, communicate to the IIPLC that a foreign antigen is present and effect changes in the IIPLC that ensure efficient loading of peptides derived from BCR-bound antigen on newly synthesized class-II molecules. In addition to the BCR, the B cell is endowed with a number of signaling receptors which allow the B cell to sense not only the presence of a foreign antigen but whether that antigen presents a potential threat to the individual, whether an adequate antibody response has already been mounted and how the B cell's interactions with helper T cells are progressing. Indeed, as discussed below, evidence is emerging that the assembly and presentation of processed antigen class-II complexes is controlled in B cells by signals transduced through several B-cell membrane proteins, including CD19/CD21, FcγRIIB1, CD40 and the class-II molecules themselves. The ability to regulate antigen processing is selective in that several other surface signaling receptors, including CD45, CD22 and tumor-necrosis-factor receptor (TNF-R) and the B-cell mitogen lipopolysaccharide (LPS), have either minimal or no demonstrable affect on APC function. In addition, antigen processing is regulated, during viral infection and in B cells during development and by induction of the stress response. The evidence for each of these will be discussed below.

## 4.1 Signaling Receptors and Their Effect on B-Cell Antigen Processing

An array of molecules capable of signal transduction has been identified on the surface of B cells. Several of these molecules have been shown to amplify or dampen the magnitude of the signal transduction cascade initiated by BCR cross-linking and in this way modulate the B cell's response to antigen. In addition, signal-transduction cascades initiated by ligands binding to B-cell signaling receptors signal independently to induce biological responses. Here, we review our current understanding of how signaling receptors modulate the APC function of B cells.

### 4.1.1 The FcγRIIB1

The B-cell receptor for IgG (FcγRIIB1) has been shown to be a negative co-receptor of the BCR (reviewed in Ravetch 1994; Daeron 1997 and in this volume). The FcγRIIB1 binds to antigen–antibody complexes and modulates B-cell activation triggered by the BCR, resulting in inhibition of antibody responses. Presumably this pathway serves as a feedback mechanism used to regulate B-cell activation when sufficient antibody has been produced to result in immune-complex formation. Immune complexes or whole Ig-specific antibodies which simultaneously cross-link the surface FcγRIIB1 with the BCR inhibit the activation of splenic B cell proliferation (Phillips and Parker 1984). The inhibition of B-cell activation is reversed in the presence of FcγRIIB-specific antibodies that prevent the binding of the Fc portions of Ig-specific antibodies to the FcγRIIB1. B-cell activation triggered through BCR cross-linking is aborted by co-ligation of the FcγRIIB1 and BCR via a mechanism that requires a 13-amino-acid sequence in the cytoplasmic domain of the FcγRIIB1 named ITIM for immunoreceptor tyrosine-based inhibition motif (Muta et al. 1994). Although the precise mechanism by which the FcγRIIB1 blocks BCR signaling is not completely understood, current evidence suggests that the ITIM of the FcγRIIB1, once phosphorylated, functions to recruit SH2-containing phosphotyrosine phosphatases to the BCR (D'AMBROSIO et al. 1995). Recent evidence indicates that FcγRIIB1 inhibition of BCR-induced B-cell activation is integrated by the dephosphorylation of CD19, leading to inactivation of PI-3 kinase activity (HIPPEN et al. 1997).

We recently investigated the effect of FcγRIIB1 on B-cell APC function and provided evidence that co-ligating of the BCR and FcγRIIB1 by whole Ig-specific antibodies regulates both signaling through the BCR for enhanced processing activity as well as the BCR-mediated internalization of antigen (WAGLE et al. 1999). Treatment of splenic B cells with $F(ab')_2$ anti-Ig but not whole anti-Ig significantly enhanced the B cell's ability to process and present antigens taken up by fluid-phase pinocytosis. Cross-linking of the FcγRIIB1 to itself had no effect on antigen processing, indicating that signaling through FcγRIIB1 did not directly affect B-cell antigen processing and that its effect is likely to be on BCR signaling. The reduced activity of the whole anti-Ig compared with $F(ab')_2$ anti-Ig was due to the interaction of anti-Ig with the FcγRIIB1, as anti-Ig treatment was as effective as $F(ab')_2$ anti-Ig when binding to the FcγRIIB1 was blocked by a FcγRIIB-specific monoclonal antibody. Moreover, anti-Ig and $F(ab')_2$ anti-Ig were equally stimulatory to a B-cell line that did not express FcγRIIB1. Thus, the FcγRIIB1 appears to indirectly regulate antigen processing by modulating the BCR-induced signals which regulate antigen-processing activity rather than directly affecting antigen processing.

The FcγRIIB1 also affects the processing of antigen which is internalized bound to the BCR. The processing and presentation of an antigen covalently coupled to anti-Ig was significantly decreased compared with the same antigen coupled to $F(ab')_2$ anti-Ig. The reduced processing activity was attributed to the interaction of the Fc region of antigen-anti-Ig conjugates with the FcγRIIB1 because the processing of antigen conjugates prepared using either the whole anti-Ig or the $F(ab')_2$ anti-Ig were similar in cells which do not express FcγRIIB1 and in splenic B cells treated with a FcγRIIB-specific monoclonal antibody to block Fc binding. Further investigation showed that internalization of antigen by B cells was reduced in the presence of whole anti-Ig compared with $F(ab')_2$ anti-Ig; however, the internalized antigen was correctly targeted to the IIPLC.

The mechanism by which the FcγRIIB1 reduces BCR internalization remains to be determined. The FcγRIIB1 expressed by B cells, unlike the FcγRIIB2 expressed by macrophages, is not an internalizing receptor, does not cluster in coated pits and tends to form large caps on cross-linking (MIETTINEN et al. 1989; AMIGORENA et al. 1992). In macrophages, the FcγRIIB2 transports antigen into the cell for subsequent processing and in this fashion provides a function analogous to the BCR in B cells (MANCA et al. 1991). In B cells, the cytoplasmic domain of the FcγRIIB1 contains an insert that increases it in length from 47 to 94 amino acids and disrupts the domain required for accumulation in coated pits (MIETTINEN et al. 1989). The FcγRIIB1, when co-ligated to the BCR, may tether the BCR to the plasma membrane and in this way block internalization. Alternatively, the effect of the FcγRIIB1 on BCR-mediated processing may be entirely due to its ability to downregulate BCR signaling. The observation described above, that the enhanced rate of internalization of the BCR on cross-linking is not dependent on kinase activity, favors the former explanation.

Taken together, these results indicate that the FcγRIIB1 is a negative regulator of the BCR-mediated antigen-processing function. Thus, in addition to the effect of the FcγRIIB1 on BCR signaling for proliferation, the FcγRIIB1 may indirectly regulate B-cell activation by regulating antigen presentation to helper T cells.

### 4.1.2 The CD19/CD21 Complex

CD19 is an essential co-receptor for the BCR which is required for the formation of germinal centers and normal antibody responses to T-cell-dependent antigens (reviewed in FEARON and CARTER 1995, TEDDER et al. 1997 and in this volume). CD19 forms a complex with the complement receptor type 2 (CD21), CD81 and Leu-13 on the B-cell membrane. The association of CD19 with CD21 allows the physiologically relevant co-ligation of CD19 with the BCR in vivo following the binding by CD21 of the C3d fragment of the C3 complement component covalently attached to antigens. Through its association with CD21, CD19 links the antigen-specific B-cell response to the immune system's innate recognition of antigen. Thus, the CD19/CD21 complex allows the B cell to identify protein antigens that have been tagged with C3d, indicating that the antigens are microbial in origin and likely represent a potential threat to the individual. The CD19/CD21 complex achieves its biological response by enhancing signaling initiated through the BCR. Co-ligation of the CD19/CD21 complex with the BCR lowers the number of BCRs which must be engaged in order to induce increases in intracellular calcium and B-cell proliferation (CARTER et al. 1991; CARTER and FEARON 1992; DEMPSEY et al. 1996).

At least two pathways of BCR signaling are enhanced by CD19: phosphatidlyinositol metabolism leading to elevated levels of intracellular $Ca^{2+}$, and the microtubule-associated-protein (MAP)-kinase cascade (CARTER et al. 1991). Mice which are deficient in CD19 show a dramatic reduction in antibody responses to T-cell-dependent antigens, but normal or augmented responses to T-cell-independent antigens (RICKERT et al. 1995). DEMPSEY et al. (1996) recently showed that linking repeated subunits of recombinant C3d to the antigen hen-egg lysozyme resulted in a potent immunogen capable of inducing antibody responses in the absence of adjuvants. Although the effect of CD19/CD21 on B-cell activation has been ascribed, in general, to its effect on BCR signaling, recent results indicate that CD21 also provides signals independent of those that influence BCR signaling. FISCHER et al. (1998) showed that CD21 expression was essential for the survival of B cells in the germinal center, even for B cells which expressed BCRs of extraordinarily high affinity. Thus, the CD19/21 complex may provide signals which independently regulate B-cell activation and survival in addition to lowering the BCR-signaling threshold of B-cell activation.

We have begun to explore the role of CD19 in the antigen-processing pathway and have documented that CD19 acts as a negative regulator of APC function (WAGLE et al. 1999). We observed that cross-linking CD19 on the B-cell surface significantly blocked processing of antigens taken up bound to the BCR but did not affect the presentation of a peptide which did not require processing. We showed that the increased rate of internalization of the BCR from the cell surface following BCR cross-linking is significantly decreased by cross-linking CD19. Moreover, we also showed that the increased rate of transport of the BCR to the IIPLC is blocked by cross-linking CD19; however, cross-linking CD19 did not affect the biosynthesis of class-II molecules or the route of trafficking of the BCR to the IIPLC. The inhibition of B-cell antigen processing by cross-linking CD19 on B-cell surfaces may be a reflection of a well-characterized biological effect attributed to CD19, namely the inhibition of B-cell activation by BCR signaling. Indeed, cross-linking of CD19 to itself has been shown to inhibit B-cell proliferation induced by BCR cross-linking (Pezzutto et al. 1987; Callard et al. 1992). This effect of CD19 may be due to its ability to recruit TAPA-1 (target of an antiproliferative antibody-1), a protein originally identified by its ability to inhibit spontaneous proliferation of transformed lymphoblastoid cell lines when cross-linked (Oren et al. 1990).

Concerning the molecular mechanism by which CD19 influences antigen processing, it is interesting to speculate that the effect of CD19 on the antigen-processing pathway is related to its recruitment of phosphatidylinositol 3 (PI-3) kinase, as suggested by Fearon and Carter (1995). We have shown previously that the PI-3 kinase inhibitor wortmannin blocks B-cell antigen processing (Song et al. 1997). The block is, at least in part, attributed to the failure to transport DM and the proteases required for Ii degradation to the IIPLC. We also identified a PI-3 kinase activity associated with the IIPLC. Thus, it is possible that CD19 signaling influences the IIPLC associated PI-3 kinase activity and thus suppresses antigen processing activity.

Thus far, we have only investigated the effect of cross-linking CD19 on B-cell antigen processing and do not know what effect co-ligation of CD19/CD21 and the BCR will have on BCR-mediated processing. Based on the synergistic effect of co-ligating CD19/CD21 and the BCR on B-cell signaling, we anticipate that the effect will be positive. We hypothesize that the involvement of the CD19/CD21 complex in B-cell processing may be critical in B-cell antigen processing in vivo. Although antigen processing by B cells is readily measurable in vitro, in the absence of CD19/CD21 involvement, the processing activity measured in vitro may be insufficient for B-cell/helper-T cell interactions in vivo. Indeed, a requirement for CD19/CD21 in BCR-mediated antigen processing in vivo is consistent with the deficit in T-cell-dependent, but not T-cell-independent antibody responses observed in CD19-deficient mice (Rickert et al. 1995) and the effectiveness of C3d-containing antigens in evoking antibody responses in vivo (Dempsey et al. 1996). Thus, the CD19/CD21 complex may serve to link antigen-specific B-cell processing to the innate immune system's recognition of antigen.

#### 4.1.3 The Class-II Molecules

In addition to the presentation of antigenic peptides to helper T cells, the class-II molecules have been demonstrated to transduce signals which regulate B-cell function (reviewed in SCHOLL and GEHA 1994; WADE et al. 1993). In conjunction with interleukin (IL)-4 and treatment with anti-Ig cross-linking, class-II molecules have been shown to activate B cells to proliferate (BALUYUT and SUBBARAO 1988; CAMBIER and LEHMANN 1989) and differentiate (BISHOP and HAUGHTON 1986). Class-II signaling has also been implicated to play a role in programmed cell death (TRUMAN et al. 1996) and to influence the co-stimulatory function of B cells by inducing the expression of important surface-adhesion and cell-interaction molecules. Indeed, cross-linking class-II molecules induces B7-1 (CD80) expression and homotypic adhesion by both leukocyte-function-associated-antigen-1 (LFA-1; CD11a/CD18)-dependent and -independent mechanisms (MOURAD et al. 1990; KANSAS et al. 1992).

Class-II signal transduction appears to involve at least two pathways that depend in part on the species of the B cells. In resting mouse B cells, cross-linking class-II molecules results in the accumulation of cyclic adenosine monophosphate (cAMP) and the translocation of protein-kinase C to the membranes (CAMBIER et al. 1991b). The transduction of the cAMP-mediated signals through class-II molecules requires the class-II cytoplasmic domain and B cells that express truncated class-II molecules lacking cytoplasmic tails are markedly impaired in their APC function, a defect that can be restored by treatment with cAMP (WADE et al. 1993). Cross-linking the class-II molecules in conjunction with BCR cross-linking in human and mouse B cells activates phospholipase C, increases intracellular calcium levels and results in the phosphorylation of a variety of cytoplasmic proteins (LANE et al. 1990; MOONEY et al. 1990; CAMBIER et al. 1991a; ANDRE et al. 1994). Superantigens which bind to class-II molecules with high affinity to a site outside the peptide binding groove have been demonstrated to activate Src-related protein kinases (MORIO et al. 1994). The domains of the class-II molecules required for signal transduction have not been clearly delineated. Current evidence suggests that both the transmembrane regions and cytoplasmic domains of the β chain participate in signal transduction and that these may have distinct signaling functions (NABAVI et al. 1989; St. PIERRE and WATTS 1990; ANDRE et al. 1994; HARTON et al. 1995). The cytoplasmic domains have been suggested to play roles in tethering the class-II molecules to the cytoskeleton and in recruiting signal-transducing proteins.

A link between class-II signaling and antigen processing and presentation was provided by the observation that class-II molecules with truncations in the cytoplasmic domains which were deficient in signaling were deficient in antigen presentation, especially to autoreactive T cells (CHEN et al. 1985; NABAVI et al. 1989; St. PIERRE and WATTS 1990; WADE et al. 1994). We investigated the effect of cross-linking class-II molecules on the ability of B cells to process and present antigen and observed that cross-linking the I-A class-II molecules on B cells using class-II-specific monoclonal antibodies significantly augmented B-cell antigen processing via the I-E molecules (FAASSEN and PIERCE 1995). This effect was only observed for the processing and presentation of antigens taken up by the B cells following binding to the BCR. Thus, cross-linking the class-II molecules had no effect on the processing and presentation of antigens taken up by fluid-phase pinocytosis or on the presentation of peptides which did not require processing. This result suggested that the augmentation of BCR-mediated antigen processing was not the result of the induction of co-stimulatory or adhesion molecules as, if this were the case, all modes of processing and presentation should be equally affected. Moreover, we showed directly that the augmentation of BCR-mediated processing following class-II cross-linking was not blocked by the inclusion of B7-specific or LFA-1-specific antibodies (FAASSEN and PIERCE 1995). Augmentation of antigen processing was enhanced by treatment of B cells with dibutyryl cAMP, a second messenger in one of the class-II signaling pathways. The cross-linking of class II does not alter the rate or number of sIg molecules internalized or the biosynthesis or expression of class-II molecules themselves.

Taken together, these results suggest that, when ligated, the class-II molecules themselves function to provide to the B-cell information that controls BCR-mediated antigen processing. The majority of studies describing the signaling function of the class-II molecules have used monoclonal antibodies as cross-linking reagents, presumably acting as a surrogate TCR. However, the natural ligand for the class-II molecules that could effect cross-linking or how cross-linking might be achieved is not known. It is possible that the TCR and CD4 engage the peptide-class-II complexes on the B-cell surface and stimulate the BCR-mediated processing, resulting in increasing numbers of peptide-class-II complexes on the B-cell surfaces for TCR engagement, increasing the strength of the T-cell/B-cell interaction. Alternatively, stimulation of BCR-mediated antigen processing could conceivably occur when superantigens bind to the class-II molecules. Superantigens have the capacity to cross-link class-II molecules (SCHOLL and GEHA 1994). In this case, superantigens would function to stimulate BCR-mediated processing, preferentially promoting antigen-specific B-cell/T-cell interactions during the T-cell responses to superantigens. An understanding of the role of class-II-molecule signaling in B-cell antigen processing will likely await a more detailed understanding of the repercussions on B cells of TCR engagement of peptide-class-II complexes.

### 4.1.4 CD40

The interaction of CD40-expressing B cells and CD154-expressing activated T cells plays a critical role in B-cell activation, differentiation and survival (reviewed in CLARK et al. 1996, GREWAL and FLAVELL 1998 and in this volume). The cross-linking of CD40 on B-cell surfaces either by CD40-specific antibodies or by soluble CD154 induces a variety of responses resulting in proliferation, isotype switching, prevention of apoptosis of germinal-center B cells and expression of costimulatory and adhesion molecules. The importance of CD40 in the induction of T-cell-dependent antibody responses in vivo has been demonstrated by the observation of impaired Ig class switching and germinal-center formation in CD40-deficient mice (KAWABE et al. 1994; XU et al. 1994) and the demonstration that the gene encoding CD154 is defective in patients with hyper-IgM syndrome (CALLARD et al. 1993; RAMESH et al. 1994).

We investigated the effect of CD40–CD154 interactions on the ability of B cells to process and present antigen and observed that ligation of CD40 on B cells by either a CD40-specific monoclonal antibody or by recombinant CD154 expressed by insect cells stimulated the processing of antigen by B cells (FAASSEN et al. 1995). CD40 ligation stimulated the processing of antigen taken up both by the BCR and by fluid-phase pinocytosis and, thus, presumably affected a step in common to the two pathways. However, CD40 ligation did not affect the presentation of peptides that did not require processing, indicating that a step in the intracellular processing pathway was the target of CD40-mediated regulation. Consistent with this observation, we found no evidence that the CD40-induced augmentation of processing is attributable to the effect of CD40 on the cell-surface expression of B7, LFA-1 and CD23. Moreover, the CD40-induced augmentation of processing was not dependent on B-cell proliferation. Thus, CD40–CD154 interactions appear to have the ability to stimulate B-cell antigen processing. This function of CD40 may play an important role in vivo allowing CD154$^{+}$ helper T cells which have been activated by APCs, such as dendritic cells, both to communicate to the antigen-specific B cell that the immune system is indeed mounting a significant response to the antigen and to stimulate the B cell's processing of antigen to insure adequate interactions with the helper T cells. The relationship between this signaling function of CD40 and those that result in B-cell proliferation and differentiation remains to be determined.

## 4.2 Signaling Receptors That Do Not Affect B-Cell Antigen Processing

The ability of the signaling receptors described above, namely the BCR, the CD19/CD21 complex, FcγRIIB1, class-II molecules and CD40, to influence the BCR-mediated processing and presentation of antigen does not appear to be the property of all membrane signaling receptors on B cells. Indeed, we were unable to demonstrate an effect on antigen processing of several other B-cell signaling receptors, including CD22, CD45, TNF-R, IL-4R and LPS-CD14. In each case, B cells or B-cell lines were treated with either the receptor's ligand or monoclonal antibodies to these receptors and tested for their ability to process and present antigens taken up either by the BCR or by fluid-phase pinocytosis or to present peptides that do not require processing. No significant effects were observed in any case. Of course, the failure to observe an effect under a particular set of conditions does not rule out the possibility that these receptors have the potential to influence B-cell processing and presentation. For stimulants such as LPS, which primarily influence B-cell proliferation, we did not expect an effect on antigen processing based on our observation that the effect of the signaling receptors described above on antigen processing was in no case dependent on B-cell proliferation. Indeed, the effects on the processing pathway occur immediately, within minutes following receptor signaling. However, for receptors such as CD45, which through its protein tyrosine phosphatase activity directly influences BCR signaling (RETH and WIENANDS 1997), we did anticipate regulation of BCR-mediated antigen processing and were surprised to observe no effect. Clearly, at this point our understanding of the potential of the BCR-mediated pathway to be regulated is incomplete.

## 4.3 Other Factors That Regulate B-Cell Antigen Processing

In addition to the signaling receptors described above that influence antigen processing by B cells, the ability of B cells to process antigen can be influenced by other factors, including the developmental stage of the B cells, the induction of the stress response in B cells and viral infection of B cells. Our current understanding of these phenomena are described below.

### 4.3.1 Regulation of Antigen Processing During B-Cell Development

During its lifetime a B cell is not equally responsive to antigen, and there are discrete developmental stages during which the engagement of antigen by the BCR results in tolerance rather than activation (reviewed in HEALY and GOODNOW 1998 and in this volume). One well-documented developmental stage at which cross-linking the BCR aborts B cells is in immature B cells soon after the expression of the BCR (METCALF et al. 1979). At this stage, B cells that express receptors with high affinity for self antigens must be eliminated. Early studies of the mechanism of immature B tolerance showed that antigen binding alone by immature B cells in the absence of helper T cells resulted in tolerance (METCALF et al. 1979). However, antigen-specific T cells could rescue the immature B cell from tolerization such that antigen binding resulted in activation. The ability of B cells to be rescued by helper T cells would be anticipated to be dependent on their ability to process and present antigen.

We investigated the ability of B cells from neonatal mice, as a source of immature B cells, to process and present antigen and found that, even though B cells from neonatal mice were adult-like in their expression of class-II molecules and sIg by 7–14 days after birth, they were unable to process and present antigen which entered the B cell bound to the BCR or by fluid-phase pinocytosis until 28 days of age. The defect appeared to be in a step in the intracellular processing pathway as neonatal B cells from 7-day-old mice were adult-like in their presentation of antigenic peptides which did not require processing. In addition, the BCR in neonatal mice functioned to bind and internalize antigen in B cells from 7-day-old mice. Thus, B cells from neonatal mice appeared to have a selective defect in their ability to process antigen and assemble peptide-class-II complex but had normal cell-surface class-II expression and a BCR which functioned to internalize antigen.

The molecular mechanism underlying antigen-processing defects in immature B cells is not known. The recent understanding of the roles of DM and DO in the assembly of peptide-class-II complexes described above suggests these as possible candidates for regulation of processing in B cells in neonatal mice. Indeed, the processing phenotype of B cells from neonatal mice, namely an inability to process native antigens but an ability to present peptides which do not require intracellular processing, bears a striking resemblance to the phenotype of DM-deficient B cells (Roche 1995). This phenotype would be predicted if the function of DM, the catalyst for peptide loading, was decreased or if the function of DO, the inhibitor of DM function, was increased in neonatal B cells. Another hallmark of a DM deficiency is the cell-surface expression of CLIP-class-II complexes. Although B cells from neonatal mice appear to be adult-like in their cell-surface expression of class-II molecules by 7–14 days after birth, we do not yet know whether the class-II molecules are filled with the normal array of peptides found bound to class-II molecules in B cells from adult mice rather than CLIP. It will be of considerable interest to characterize the expression of DM and DO in immature B cells.

Concerning the possible function of reduced processing activity in immature B cells, we speculate that, during developmental stages in which self-reactive B cells must be eliminated, their antigen processing function may be decreased in order to prevent rescue from tolerance by interactions with antigen-specific helper T cells. In this context, the observation of a defect in antigen processing in neonatal B cells raises several questions. Are all tolerance-susceptible B cells deficient in their antigen-processing function? For example, are B cells in germinal centers which have undergone somatic mutation and are tolerance-sensitive capable of processing antigen? Are immature B cells in the bone marrow of adult individuals capable of processing and presenting antigen? Can the antigen-processing function in neonatal B cells be stimulated by any of the signaling receptors that regulate antigen processing in adult cells? Following antigen binding, do neonatal B cells avoid interactions with antigen-specific T cells, unless provided with additional information indicating that the B cell's participation in the immune response would be beneficial? Clearly, answers to these questions will increase our understanding of tolerance mechanisms during B-cell development.

### 4.3.2 The Stress Response and Antigen Processing

The heat-shock response is a universal and highly conserved cellular response to a variety of unfavorable conditions or environmental stresses (Gething and Sambrook 1992; Craig et al. 1993). The stress response was first observed as a reaction to elevated temperatures, and thus the term heat-shock response is used synonymously with stress response. Other stress-inducing agents known to elicit a similar cellular response include oxidants such as hydrogen peroxide, γ-radiation, sodium arsenite, ethanol, amino-acid analogues, heavy metals, calcium ionophores and various agents that inhibit glycoprotein processing (Craig et al. 1993). The stress response is induced by a variety of factors, many of which are highly relevant to immune responses (Polla 1988). These include fever, bacterial and viral infections, oxidant injury and inflammation. Thus, the stress response has the potential to communicate to the immune system that a dangerous infection or injury has occurred.

The stress response culminates in the increased synthesis of a number of proteins termed heat-shock proteins (hsps; Gething and Sambrook 1992). Although the expression of certain hsps is greatly induced by stress, it is now clear that cells have large quantities of constitutively expressed hsps and that these hsps are involved in all aspects of protein folding and oligermerization. In addition, hsps function in the intracellular transport of folding intermediates to appropriate subcellular sites, the disassembly of oligomeric structures and the removal of aggregated or misfolded proteins. The term molecular chaperones has been used to describe the involvement of the hsps in these processes.

Another important aspect of the cellular response to stress-inducing agents is the induction of the stress-activated protein kinase (SAPK), otherwise known as the c-Jun NH2-terminal kinases (JNKs; Keyse and Emslie 1992; Kyriakis et al. 1994; Derijard et al. 1994). These protein tyrosine kinases were discovered to form a cascade, which is parallel to, but distinct from, the MAP-kinase pathway of signal transduction. The SAPKs ultimately activate the transcriptional activator c-Jun, which stimulates transcription of a wide variety of genes. The relationship between the SAPK pathway and the induction of hsp synthesis is not known.

Class-II-restricted antigen processing has been shown to be affected by stress. Several studies have reported that heat-shock treatment of B cells results in an enhanced ability to process and present antigens in the class-II pathway. In each of these cases, a concomitant increase in synthesis of hsps is observed, although no direct link between these hsps and antigen processing has been found, nor is it known whether signaling through the SAPK pathway is required for the effect. In studies on human Epstein-Barr virus, transformed B cells (Rees et al. 1991), treatment of cells at 44°C for short durations prior to the addition of exogenous antigen caused an increase in processing of whole ultraviolet (UV)-attenuated influenza virus as well as presentation of a hemagglutinin peptide. This was accompanied by a modest increase in cell-surface class-II expression, which could not fully account for the observed increase in T-cell stimulation. The effects of hydrogen peroxide, another inducer of the stress response, were also assessed and were found to augment peptide presentation to an extent greater than that seen with heat shock. Studies of a mouse B lymphoblastoid cell line showed that heat shock caused an increase in class-II presentation of endogenously synthesized chicken ovalbumin (Michalek et al. 1992), but not exogenously added ovalbumin or beef insulin. In this study, the heat-shock treatment preceded antigen administration. Our own experiments on a mouse B lymphoma demonstrated that heat shock only affected antigen processing if the cells were exposed to elevated temperature when antigen was given (Cristau et al. 1994). Here, dramatic upregulation was seen on fluid-phase and receptor-mediated antigen processing, as well as on the presentation of peptide antigen which did not require processing. The increase in processing activity was accompanied by induction of members of the hsp70 and hsp90 families and could not be attributed to increase in cell-surface expression of class-II molecules. Furthermore, the class-II molecules purified from heat-shocked cells were better able to stimulate T cells in vitro than class-II molecules from untreated cells. This enhanced stimulatory ability was attributed either to a difference in the population of peptides which co-purify with the class-II molecules or to a difference in the post-translational modification of the heat-shocked class-II molecules.

In a variety of experimental models, stress has been implicated in initiating or sustaining autoimmune or inflammatory responses, possibly by influencing the array of peptides bound to the class-II molecules (POLLA 1988). We showed that the acquisition of peptide by newly synthesized class-II is accelerated during heat shock, which was accompanied by an increase in the rate of invariant chain degradation (CRISTAU et al. 1994). Recent analysis by means of high-performance liquid chromatography revealed that the overall quantity and quality of peptides bound to class-II peptides is not significantly altered by heat shock. However, there was a small increase in the amount of class-II-associated peptide derived from the 70-kDa heat-shock cognate protein following heat shock. The rate of class-II egress from the Golgi is not affected by heat shock, nor is the subcellular site of antigenic peptide binding to class-II molecules.

Taken together, these results indicate that following the induction of the stress response, B cells accelerate the rate of assembly of peptide class-II complexes, but the site of assembly and the array of peptides bound to the class-II molecules does not change significantly. It may be that the accelerated rate of assembly has a significant impact on the ability of B cells to acquire T cell help in vivo.

### 4.3.3 The Effect of Virus Infection

Studies carried out in our laboratory documented an effect of viral infection on B-cell APC function (DOMANICO and PIERCE 1992). Influenza-virus and vaccinia-virus infection blocked processing and presentation of antigen by B cells. This block appeared to be primarily within the processing pathway, as viral infection had little effect on presentation of peptide. Only live infectious virus, not UV-irradiated virus, blocked the processing function, indicating that there was no competition of viral particles with antigen for the class-II-processing machinery. The mechanisms by which these very different viruses affect B-cell antigen processing remain to be determined. Recent studies of the class-I-processing pathway have revealed a variety of mechanisms by which specific viral-gene products function to block discrete steps in the pathway (FRUH et al. 1998). Whether the effects of the influenza virus and vaccinia virus are due to specific effects rather than to general effects on cellular function remains to be determined. It will also be of interest to determine whether viruses for which B cells are the natural hosts influence antigen processing.

## 5 Conclusions

The ability of a B cell to assemble antigenic peptide class-II complexes is key to its ability to interact with antigen-specific helper T cells and, as such, plays a central role in the initiation of antibody responses. Current evidence indicates that this assembly process is an important target of regulation, presumably serving to control B-cell/T-cell interactions. The principle regulator of B-cell antigen processing appears to be the BCR itself, which endows the B cell with the ability to discriminate foreign versus self antigens and ensures that foreign antigens are preferentially processed and presented. The functions of the BCR in antigen processing are themselves targets of regulations by a variety of internal and external signals that provide information to the B cell concerning the appropriateness of the response. When integrated, this information results in the appropriate modulation of the assembly of peptide class-II complexes. Clearly, the ability of a B cell to correctly interpret all information provided to it and to regulate antigen processing and presentation accordingly has important implications for the initiation of beneficial versus autoimmune responses. At present, molecular targets of regulation in the antigen-processing pathway remain to be elucidated. Hopefully, a clearer picture of how this essential function of the B cell is controlled will develop as our understanding of the cellular and molecular mechanisms underlying the BCR-mediated antigen-processing pathway expands.

## References

Aluvihare VR, Khamlichi AA, Williams GT, Adorini L, Neuberger MS (1997) Acceleration of intracellular targeting of antigen by the B-cell antigen receptor: importance depends on the nature of the antigen-antibody interaction. EMBO J 16:3553–3562

Amigorena S, Bonnerot C, Drake JR, Choquet D, Hunziker W, Guillet JG, Webster P, Sautes C, Mellman I, Fridman WH (1992) Cytoplasmic domain heterogeneity and functions of IgG Fc receptors in B lymphocytes. Science 256:1808–1812

Anderson M, Miller J (1992) Invariant chain can function as a chaperone protein for class II major histocompatability complex molecules. Proc Natl Acad Sci USA 89:2282–2286

Andre P, Cambier JC, Wade TK, Raetz T, Wade WF (1994) Distinct structural compartmentalization of the signal transducing functions of major histocompatibility complex class II (Ia) molecules. J Exp Med 179:763–768

Baluyut AR, Subbarao B (1988) The synergistic effects of anti-IgM and monoclonal anti-Ia antibodies in induction of murine B lymphocyte activation. J Mol Cell Immunol 4:45–57

Bishop GA, Haughton G (1986) Induced differentiation of a transformed clone of Ly-1+ B cells by clonal T cells and antigen. Proc Natl Acad Sci USA 83:7410–7414

Bonnerot C, Lankar D, Hanau D, Spehner D, Davoust J, Salamero J, Fridman WH (1995) Role of B cell receptor Igα and Igβ subunits in MHC class II-restricted antigen presentation. Immunity 3:335–347

Calafat J, Nijenhuis M, Janssen H, Tulp A, Dussseljee S, Wubbolts R, Neefjes J (1994) Major histocompatibility complex class II molecules induce the formation of endocytic MIIC-like structures. J Cell Biol 126:967–977

Callard RE, Rigley KP, Smith SH, Thurstan S, Shields JG (1992) CD19 regulation of human B cell responses. B cell proliferation and antibody secretion are inhibited or enhanced by ligation of the CD19 surface glycoprotein depending on the stimulating signal used. J Immunol 148:2983–2987

Callard RE, Armitage RJ, Fanslow W, Spriggs MK (1993) CD40 ligand and its role in X-linked hyper-IgM syndrome. Immunol Today 14:559–564

Cambier JC, Lehmann KR (1989) Ia-mediated signal transduction leads to proliferation of primed B lymphocytes. J Exp Med 170:877–886

Cambier JC, Morrison DC, Chien MM, Lehmann KR (1991a) Modeling of T cell contact-dependent B cell activation: IL-4 and antigen receptor ligation primes quiescent B cells to mobilize calcium in response to Ia cross-linking. J Immunol 146:2075–2082

Cambier JC, Newell MK, Justement LB, McGuire JC, Leach KL, Chen ZZ (1991b) Ia binding ligands and cAMP stimulate nuclear translocation of PKC in B lymphocytes. Nature 327:629–632

Carter RH, Tuveson DA, Park DJ, Rhee SG, Fearon DT (1991) The CD19 complex of B lymphocytes: activation of phospholipase C by a protein tyrosine kinase-dependent pathway that can be enhanced by the membrane IgM complex. J Immunol 147:3663–3671

Carter RH, Fearon DT (1992) CD19: Lowering the threshold for antigen receptor stimulation of B lymphocytes. Science 256:105–107

Cassard S, Salamero J, Daniel H, Spehner D, Davoust J, Fridman WH, Bonnerot C (1998) A Tyrosine-Based Signal Present in Igα Mediates B Cell Receptor Constitutive Internalization. J Immunol 160:1767–1773

Castellino F, Germain R (1995) Extensive trafficking of MHC class II-invariant chain complexes in the endocytic pathway and appearance of peptide-loaded class II in multiple compartments. Immunity 2:73–88

Casten LA, Lakey EK, Jelachich ML, Margoliash E, Pierce SK (1985) Anti-immunoglobulin augments the B-cell antigen-presentation function independently of internalization of receptor-antigen complex. Proc Natl Acad Sci USA 82:5890–5894

Casten LA, Pierce SK (1988) Receptor-mediated B cell antigen processing: increased antigenicity of a globular protein covalently coupled to antibodies specific for B cell surface structures. J Immunol 140:404–410

Chen JW, Murphy TL, Willingham MC, Pastan I, August JT (1985) Identification of two lysosomal membrane glycoproteins. J Cell Biol 101:85–95

Cheng PC, Steele CR, Gu L, Song W, Pierce SK (1999) MHC class II antigen processing in B cells: accelerated intracellular targeting of antigens. J Immunol (in press)

Clark LB, Foy TM, Noelle RJ (1996) CD40 and its ligand. Adv Immunol 63:43–78

Craig EA, Gambill BD, Nelson RJ (1993) Heat shock proteins: molecular chaperones of protein biogenesis. [Review] Microbiol Rev 57:402–414

Cresswell P (1994) Assembly, transport function of MHC class II molecules. Ann Rev Immunol 12: 259–295

Cristau B, Schafer PH, Pierce SK (1994) Heat shock enhances antigen processing and accelerates the formation of compact class II ab dimers. J Immunol 152:1546–1556

D'Ambrosio D, Hippen KL, Minskoff SA, Mellman I, Pani G, Siminovitch KA, Cambier JC (1995) Recruitment and activation of PTP1 C in negative regulation of antigen receptor signaling by FcgRIIB1. Science 268:293–297

Daeron M (1997) Fc Receptor Biology. Ann Rev Immunol 15:203–234

Dempsey PW, Allison MED, Akkaraju S, Goodnow CC, Fearon DT (1996) C3d of complement as a molecular adjuvant: bridging innate and acquired immunity. Science 271:348–350

Denzin LK, Sant'Angelo DB, Hammond C, Surman MJ, Cresswell P (1997) Negative regulation by HLA-DO of MHC class II-restricted antigen processing. Science 278:106–109

Derijard B, Hibi M, Wu IH, Barrett T, Su B, Deng T, Karin M, Davis RJ (1994) JNK1: a protein kinase stimulated by UV light and Ha-Ras that binds and phosphorylates the c-Jun activation domain. Cell 76:1025–1037

Domanico SZ, Pierce SK (1992) Virus infection blocks the processing and presentation of exogenous antigen with the major histocompatibility complex class II molecules. Eur J Immunol 22:2055–2062

Escola JM, Grivel JC, Chavrier P, Gorvel JP (1995) Different endocytic compartments are involved in the tight association of class II molecules with processed hen egg lysozyme and ribonuclease A in B cells. J Cell Sci 108:2337–2345

Faassen AE, Dalke D, Berton MT, Warren WD, Pierce SK (1995) CD40-CD40-ligand interactions stimulate B cell antigen processing. Eur J Immunol 25:3249–3255

Faassen AE, Pierce SK (1995) Cross-linking cell surface class II molecules stimulates Ig-mediated B cell antigen processing. J Immunol 155:1737–1745

Fearon DT, Carter RH (1995) The CD19/CR2/TAPA-1 complex of B lymphocytes: Linking natural to acquired immunity. Ann Rev Immunol 13:127–149

Fischer MB, Goerg S, Shen L, Prodeus AP, Goodnow CC, Kelsoe G, Carroll MC (1998) Dependence of germinal center B cells on expression of CD21/CD35 for survival. Science 280:582–585

Fruh K, Ahn K, Peterson PA (1998) Inhibition of MHC class I antigen presentation by viral proteins. J Mol Med 75:18–27

Germain RN (1994) MHC-dependent antigen processing and peptide presentation: providing ligands for T lymphocyte activation. Cell 76:287–299

Gething M-J, Sambrook J (1992) Protein folding in the cell. Nature 355:33–45

Glickman JN, Morton PA, Slot JW, Kornfeld S, Geuze HJ (1996) The biogenesis of the MHC class II compartment in human I-cell disease B lymphoblasts. J Cell Biol 132:769–785

Grewal IS, Flavell RA (1998) CD40 and CD154 in cell-mediated immunity. Ann Rev Immunol 16: 111–135
Griffin JP, Chu R, Harding CV (1997) Early endosomes and a late endocytic compartment generate different peptide-class II MHC complexes via distinct processing mechanisms. J Immunol 158: 1523–3152
Grupp SA, Campbell K, Mitchell RN, Cambier JC, Abbas AK (1993) Signaling-defective mutants of the B lymphocyte antigen receptor fail to associate with Ig-alpha and Ig-beta/gamma. J Biol Chem 268:25776–25779
Harton JA, Van Hagen AE, Bishop GA (1995) The cytoplasmic and transmembrane domains of MHC class II β chains deliver distinct signals required for MHC class II-mediated B cell activation. Immunity 3:349–358
Healy JI, Goodnow CC (1998) Positive versus negative signaling by lymphocyte antigen receptors. Annu Rev Immunol 16:645–670
Hippen KL, Buhl AM, D'Ambrosio D, Nakamura K, Persin C, Cambier JC (1997) FcγRIIB1 inhibition of BCR-mediated phosphoinositide hydrolysis and $Ca^{2+}$ mobilization is integrated by CD19 dephosphorylation. Immunity 7:49–58
Jiang W, Swiggard WJ, Heufler C, Peng M, Mirza A, Steinman RM, Nussenzweig MC (1995) The receptor DEC-205 expressed by dendritic cells and thymic epithelial cells is involved in antigen processing. Nature 375:151–155
Kansas GS, Cambier JC, Tedder TF (1992) CD4 binding to major histocompatibility complex class II antigens induces LFA-1-dependent and -independent homotypic adhesion of B lymphocytes. Eur J Immunol 22:147–152
Kawabe T, Naka T, Yoshida K, Tanaka T, Fujiwara H, Suematsu S, Yoshida N, Kishimoto T, Kikutani H (1994) The immune responses in CD40-deficient mice: impaired immunoglobulin class switching and germinal center formation. Immunity 1:167–178
Keyse SM, Emslie EA (1992) Oxidative stress and heat shock induce a human gene encoding a protein-tyrosine phosphatase. Nature 359:644–647
Kleijmeer MJ, Morkowski S, Griffith JM, Rudensky AY, Geuze HJ (1997) Major histocompatibility complex class II compartments in human and mouse B lymphoblasts represent conventional endocytic compartments. J Cell Biol 139:639–649
Knight AM, Lucocq JM, Prescott AR, Ponnambalam S, Watts C (1997) Antigen endocytosis and presentation mediated by human membrane IgG1 in the absence of the Igα/Igβ dimer. EMBO J 16: 3842–3850
Koch N, Koch S, Hammering GJ (1982) Ia invariant chain detected on lymphocyte surfaces by monoclonal antibody. Nature 299:644–645
Kyriakis JM, Banerjee P, Nikolakaki E, Dai T, Rubie EA, Ahmad MF, Avruch J, Woodgett JR (1994) The stress-activated protein kinase subfamily of c-Jun kinases. Nature 369:156–160
Lane PJ, McConnell FM, Schieven GL, Clark EA, Ledbetter JA (1990) The role of class II molecules in human B cell activation: association with phosphatidyl inositol turnover, protein tyrosine phosphorylation, and proliferation. J Immunol 144:384–392
Lanzavecchia A (1990) Receptor-mediated antigen uptake and its effect on antigen presentation to class II-restricted T lymphocytes. Ann Rev Immunol 8:773–793
Liljedahl M, Kuwana T, Fung-Leung W, Jackson MR, Peterson PA, Karlsson L (1996) HLA-DO is a lysosomal resident which requires association with HLA-DM for efficient intracellular transport. EMBO J 15:4817–4824
Liu S, Marks MS, Brodsky FM (1998) A dominant-negative clathrin mutant differentially affects trafficking of molecules with distinct sorting motifs in the class II major histocompatibility complex (MHC) pathway. J Cell Biol 140:1023–1037
Lotteau V, Teyton L, Peleraux A, Nilsson T, Karlsson L, Schmid SL, Quaranta V, Peterson PA (1990) Intracellular transport of class II MHC molecules directed by invariant chain. Nature 348:600–605
Manca F, Fenoglio D, LiPira G, Kunkl A, Celada F (1991) Effect of antigen/antibody ratio on macrophage update, processing, and presentation to T cells of antigen complexes with polyclonal antibodies. J Exp Med 173:37–48
Metcalf ES, Schrater F, Klinman NR (1979) Murine models of tolerance induction in developing and mature B cells. Immunol Rev 43:143–183
Michalek MT, Benacerraf B, Rock KL (1992) The class II MHC-restricted presentation of endogenously synthesized ovalbumin displays clonal variation, requires endosomal/lysosomal processing, and is upregulated by heat shock. J Immunol 148:1016–1024

Miettinen HM, Rose JK, Mellman I (1989) Fc receptor isoforms exhibit distinct abilities for coated pit localization as a result of cytoplasmic domain heterogeneity. Cell 58:317–327
Mitchell RN, Barnes KA, Grupp SA, Sanchez M, Misulovin Z, Nussenzweig C, Abbas AK (1995) Intracellular targeting of antigens internalized by membrane immunoglobulin in B lymphocytes. J Exp Med 181:1705–1714
Mooney NA, Grillot-Courvalin C, Hivroz C, Ju L, Charron D (1990) Early biochemical events after MHC class II-mediated signaling on human B lymphocytes. J Immunol 145:2070–2076
Morio T, Geha RS, Chatila TA (1994) Engagement of MHC class II molecules by staphylococcal super antigens activates src-type protein tyrosine kinases. Eur J Immunol 24:651–658
Mourad W, Geha RS, Chatila T (1990) Engagement of major histocompatibility complex class II molecules induces sustained, lymphocyte function-associated molecule 1-dependent cell adhesion. J Exp Med 172:1513–1516
Muta T, Kurosaki T, Misulovin Z, Sanchez M, Nussenzweig MC, Ravetch JV (1994) A 13-amino acid motif in the cytoplasmic domain of FcgRIIB modulates B-cell receptor signalling. Nature 368:70–73
Nabavi N, Ghogawala Z, Myer A, Griffith IJ, Wade WF, Chen ZZ, McKean DJ, Glimcher LH (1989) Antigen presentation abrogated in cells expressing truncated Ia molecules. J Immunol 142:1444–1447
Neefjes JJ, Stollorz V, Peters PJ, Geuze HJ, Ploegh HL (1990) The biosynthetic pathway of MHC class II but not class I molecules intersects the endocytic route. Cell 61:171–183
Odorizzi CG, Trowbridge IS, Xue L, Hopkins CR, Davis CD, Collawn JF (1994) Sorting Signals in the MHC class II invariant chain cytoplasmic tail and transmembrane region determine trafficking to an endocytic processing compartment. J Cell Biol 126:317–330
Oren R, Takahashi S, Doss C, Levy R, Levy S (1990) TAPA-1, the target of an antiproliferative antibody, defines a new family of transmembrane proteins. Mol Cell Biol 10:4007–4015
Parker DC (1993) T cell-dependent B cell activation. Ann Rev Immunol 11:331–360
Patel KJ, Neuberger MS (1993) Antigen presentation by the B cell antigen receptor is driven by the α/β sheath and occurs independently of its cytoplasmic tyrosines. Cell 74:939–946
Peters PJ, Neefjes JJ, Oorschot V, Ploegh H, Geuze HJ (1991) Segregation of MHC class II molecules from MHC class I molecules in the Golgi complex for transport to lysosomal compartments. Nature 349:669–676
Peters PJ, Raposo G, Neefjes JJ, Oorschot V, Leyendekker RL, Geuze HJ, Ploegh HL (1995) Major histocompatibility complex class II compartments in human B-lymphoblastoid cells are distinct from early endosomes. J Exp Med 182:325–334
Pezzutto A, Dorken B, Rabinovitch PS, Ledbetter JA, Moldenhauer G, Clark EA (1987) CD19 monoclonal antibody HD37 inhibits anti-immunoglobulin-induced B cell activation and proliferation. J Immunol 138:2793–2799
Phillips NE, Parker DC (1984) Cross-linking of B lymphocyte Fcγ receptors and membrane immunoglobulin inhibits anti-immunoglobulin-induced blastogenesis. J Immunol 132:627–632
Pleiman CM, D'Ambrosia D, Cambier JC (1994) The B-cell antigen receptor complex: structure and signal transduction. Immunol Today 15:393–399
Polla BS (1988) A role for heat shock proteins in inflammation? Immunol Today 9:134–137
Qiu Y, Xu X, Wandinger-Ness A, Dalke DP, Pierce SK (1994) Separation of subcellular compartments containing distinct functional forms of MHC class II. J Cell Biol 125:595–605
Ramesh N, Fuleihan R, Geha R (1994) Molecular pathology of X-linked immunoglobulin deficiency with normal or elevated IgM. Immunol Rev 138:87–104
Ravetch JV (1994) Fc receptors: rubor redux. Cell 78:553–560
Rees ADM, Donati Y, Lombardi G, Lamb J, Polla B, Lechler R (1991) Stress-induced modulation of antigen-presenting cell function. Immunology 74:386–392
Reth M (1992) Antigen receptor on B lymphocytes. Annu Rev Immunol 10:97–121
Reth M, Wienands J (1997) Initiation and processing of signals from the B cell antigen receptor. Ann Rev Immunol 15:453–479
Rickert RC, Rajewsky K, Roes J (1995) Impairment of T-cell dependent B-cell responses and B-1 cell development in CD19-deficient mice. Nature 376:352–356
Roche PA, Cresswell P (1990) Invariant chain association with HLA-DR molecules inhibits immunogenic peptide binding. Nature 345:615–618
Roche PA, Teletski CL, Stang E, Bakke O, Long EO (1993) Cell surface HLA-DR-invariant chain complexes are targeted to endosomes by rapid internalization. Proc Natl Acad Sci USA 90:8581–8585
Roche PA (1995) HLA-DM: An in vivo facilitator of MHC class II peptide loading. Immunity 3:259–262
Rudensky AY, Maric M, Eastman S, Shoemaker L, DeRoos PC, Blum JS (1994) Intracellular assembly and transport of endogenous peptide-MHC class II complexes. Immunity 1:585–594

Sadasivan BK, Cariappa A, Waneck GL, Cresswell P (1995) Assembly, peptide loading, and transport of MHC class I molecules in a calnexin-negative cell line. Cold Spring Harbor Symp 55:267–275
Sallusto F, Cella M, Danieli C, Lanzavecchia A (1995) Dendritic cells use macropinocytosis and the mannose receptor to concentrate macromolecules in the major histocompatibility complex class II compartment: downregulation by cytokines and bacterial products. J Exp Med 182:389–400
Sanderson F, Kleijmeer MJ, Kelly A, Verwoerd D, Tulp A, Neefjes JJ, Geuze HJ, Trowsdale J (1994) Accumulation of HLA-DM, a regulator of antigen presentation, in MHC class II compartments. Science 266:1566–1573
Sanderson F, Thomas C, Neefjes J, Trowsdale J (1996) Association between HLA-DM and HLA-DR in vivo. Immunity 4:87–96
Schafer PH, Green JM, Malapati S, Gu L, Pierce SK (1996) HLA–DM is present in one-fifth the amount of HLA-DR in the class II peptide-loading compartment where it associates with leupeptin-induced peptide (LIP)-HLA-DR. J Immunology 157:5487–5495
Scholl PR, Geha RS (1994) MHC class II signaling in B cell activation. Immunol Today 15:418–422
Song W, Cho H, Cheng P, Pierce SK (1995) Entry of the B-cell antigen receptor and antigen into the class II peptide-loading compartment is independent of receptor cross-linking. J Immunol 155:4255–4263
Song W, Wagle NW, Banh T, Whiteford CC, Ulug E, Pierce SK (1997) Wortmannin, a phosphatidylinositol 3-kinase inhibitor, blocks the assembly of peptide-MHC class II complexes. Int Immunol 9:1709–1722
St-Pierre Y, Watts TH (1990) MHC class II-restricted presentation of native protein antigen by B cells is inhibitable by cycloheximide and brefeldin A. J Immunol 145:812–818
Tedder TF, Inaoki M, Sato S (1997) The CD19-CD21 complex regulates signal transduction thresholds governing humor immunity and autoimmunity. Immunity 6:107–118
Teyton L, O'Sullivan D, Dickson PW, Lotteau V, Sette A, Fink P, Peterson PA (1990) Invariant chain distinguishes between the exogenous and endogenous antigen presentation pathways. Nature 348: 39–44
Truman JP, Choqueux C, Charron D, Mooney N (1996) HLA class II molecule signal transduction leads to either apoptosis or activation via two different pathways. Cell Immunol 172:149–157
Tulp A, Verwoerd D, Dobberstein B, Ploegh HL, Pieters J (1994) Isolation and characterization of the intracellular MHC class II compartment. Nature 369:120–126
Wade WF, Davoust J, Salamero J, André P, Watts TH, Cambier JC (1993) Structural compartmentalization of MHC class II signaling function. Immunol Today 14:539–546
Wade WF, Ward ED, Rosloniec EF, Barisas BG, Freed JH (1994) Truncation of the Aα chain of MHC class II molecules results in inefficient antigen presentation to antigen-specific T cells. Int Immunol 6:1457–1465
Wagle NM, Faassen AE, Kim JH, Pierce SK (1999) Regulation of B cell receptor-mediated MHC class II antigen processing by FcγRIIB1. J Immunol 162:2732–2740
Wagle NM, Kim JH, Pierce SK (1998) Signaling through the B cell antigen receptor regulates discrete steps in the antigen processing pathway. Cell Immunol 184:1–11
Wagle NM, Kim JH, Pierce SK (1999) CD19 regulates B cell antigen receptor-mediated MHC class II antigen processing vaccine (in press)
Watts C (1997) Capture and processing of exogenous antigens for presentation on MHC molecules. Ann Rev Immunol 15:821–850
Weber DA, Brian DE, Jensen PE (1996) Enhanced dissociation of HLA-DR-bound peptides in the presence of HLA-DM. Science 274:618–620
West MA, Lucocq JM, Watts C (1994) Antigen processing and class II MHC peptide-loading compartments in human B-lymphoblastoid cells. Nature 369:147–150
Xu J, Foy TM, Laman JD, Elliott EA, Dunn JJ, Waldschmidt TJ, Elsemore J, Noelle RJ, Flavell RA (1994) Mice deficient for the CD40 ligand. Immunity 1:423–431
Xu X, Pierce SK (1995) The novelty of antigen processing compartments. J Immunol 155:1652–1654
Xu X, Song W, Cho H, Qiu Y, Pierce SK (1995) Intracellular transport of invariant chain – MHC class II complexes to the peptide-loading compartment. J Immunol 155:2984–2992
Xu X, Press B, Wagle NM, Cho H, Wandinger-Ness A, Pierce SK (1996) B cell antigen receptor signaling links biochemical changes in the class II peptide-loading compartment to enhanced processing. Int Immunol 8:1867–1876

# Molecular Processes that Regulate Class Switching

J. Stavnezer



Department of Molecular Genetics and Microbiology and Program in Immunology and Virology, University of Massachusetts Medical School, Worcester, MA 01655-0122, USA

# 1 Introduction

Upon antigen activation, surface immunoglobulin (sIg) $M^+IgD^+$ B cells switch to express a different $C_H$ region, resulting in a change of the antibody class synthesized and a change in the effector function of the antibody produced. Class switching does not result in a change in the expressed heavy-chain V(D)J region or expressed light chain; thus, antigen-binding specificity is unaltered. The change in antibody class is effected by a deletional DNA recombination event called class-switch recombination (CSR), which occurs between tandemly repeated sequences within switch (S) regions located upstream, or 5′, of each of the $C_H$ genes (Fig. 1). In addition to direct switching from expression of IgM to expression of any of the downstream $C_H$ genes, sequential switching between downstream $C_H$ genes also occurs (reviewed in STAVNEZER 1996a).

Different Ig classes are found in most vertebrates examined, suggesting the importance of having different effector functions. There are five classes of immunoglobulins in mice, rats and humans: IgM, IgD, IgG, IgE and IgA. There are four subclasses of IgG (IgG1, IgG2b, IgG2a and IgG3 in the mouse: IgG1, IgG2, IgG3

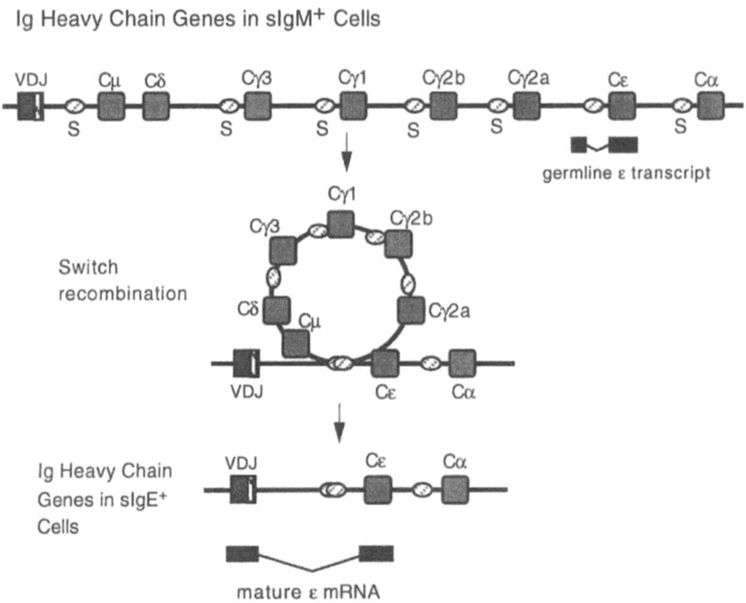


**Fig. 1.** Class-switch recombination, illustration of switching to immunoglobulin (Ig) E and germline ε transcripts. Diagram of Ig heavy-chain locus in a mouse B cell expressing IgM. *Below* is a simplified splice diagram of germline ε transcripts. As true for all germline transcripts, the primary transcript is spliced to generate a polyadenylated RNA containing an upstream or 5′ exon (I exon) spliced to the Cε exons at the normal splice acceptor for mature heavy-chain mRNA. During switch recombination, the segment of the chromosome residing between the switch (S) regions that undergo recombination is excised as a circle (see text). After switch recombination, the V(D)J segment (heavy-chain variable region) is unchanged; thus, antigen specificity is unchanged

and IgG4 in human) and, in humans, there are two subclasses of IgA (IgA1 and IgA2). The IgG subclasses of mouse and human do not show a correspondence, suggesting that the Cγ genes duplicated subsequent to the evolutionary divergence of mice and humans.

IgD expression does not require CSR, as IgD is co-expressed with IgM in mature, naive, resting B cells by alternative RNA processing of a common primary transcript (MAKI et al. 1981; MOORE et al. 1981; KNAPP et al. 1982; YUAN and TUCKER 1984; YUAN 1986; YUAN and WITTE 1988). After activation, B cells downregulate IgD expression by regulation of termination of transcription prior to the Cδ exons (YUAN and TUCKER 1984). Although the subject of whether other isotypes can be expressed by alternative RNA processing has been controversial, altogether, the data indicate that the vast bulk of switching to isotypes other than IgD occurs by DNA recombination rather than by RNA processing (CHU et al. 1993; IRSCH et al. 1994; reviewed in STAVNEZER 1996a).

$IgD^+$ memory cells are found in human tonsillar germinal centers, identified by the presence of numerous somatic mutations in their expressed $V_H$ genes (LIU et al. 1996a). These $IgD^+$ memory B cells have switched to IgD expression by deletion of their Cμ gene via a recombination event, which can occur between a few different sequences located upstream, downstream and within Sμ, with sequences located upstream of the Cδ gene that are not typical S-region sequences (GILLIAM et al. 1984; YASUI et al. 1989; OWENS et al. 1991; WHITE et al. 1990; KLUIN et al. 1996). Nothing is known about the regulation of this process. The function of IgD is controversial, and mice with targeted deletions of Cδ exons do not have a major impairment in their immune response, although they have a delay in their production of high-affinity antibodies (NITSCHKE et al. 1993; ROES and RAJEWSKY 1993).

# 2 Induction of Class Switch Recombination (CSR) and Regulation by Cytokines

## 2.1 In Vivo Induction of CSR

Since switch recombination occurs in mature B cells upon exposure to antigen, the heavy-chain class chosen is influenced by the antigen. Naive B cells have the potential to switch to any of the downstream isotypes prior to exposure to antigen and T cells or mitogens (GEARHART et al. 1975; LAYTON et al. 1984; TEALE and KLINMAN 1984; BERGSTEDT-LINDQVIST et al. 1988; LEBMAN and COFFMAN 1988; SAVELKOUL et al. 1988; reviewed by COFFMAN et al. 1993 and STAVNEZER 1996a). In vivo, during T-dependent (TD) responses, a complex of antigen with natural IgM (the non-specific and polyreactive IgM present in serum in unimmunized animals) and complement products activates B cells via the sIg receptor and co-receptor molecules, along with contact-dependent signals from T cells and signals from cytokines (COOKE et al. 1994; FEARON and LOCKSLEY 1996; O'ROURKE et al.

1997; Carroll 1998). It is likely that antigen upregulates co-stimulatory signals on B cells, thereby stimulating helper T ($T_H$) cells, but antigen also activates B cells directly by cross-linking sIg, providing a signal to the B cell in addition to the CD40 signal (Goroff et al. 1991; Cooke et al. 1994; Thornton et al. 1994). This combination of signals induces B-cell proliferation and heavy-chain class switching, along with antibody secretion and formation of memory B cells. T-independent (TI) antigens also induce class switching by delivering signals via sIg cross-linking in conjunction with signals from cytokines and possibly contact-dependent signals from a variety of cell types.

## 2.2 CD40 Signaling Can Induce CSR

The most important component of contact-dependent T-cell help for B cells is the CD40 ligand (CD40L/gp39/CD154), a 33-kDa glycoprotein, which is transiently expressed on activated $T_H$ cells and provides a mitogenic signal to B cells (Fanslow et al. 1992; Lane et al. 1992; Noelle et al. 1992a; Lane et al. 1993; reviewed by Foy et al. 1996 and van Kooten and Banchereau 1996). Humans lacking CD40L have hyper-IgM syndrome, thus named because they express large amounts of serum IgM and do not express other Ig classes except small amounts of IgG3 (Allen et al. 1993; Aruffo et al. 1993; DiSanto et al. 1993; Fuleihan et al. 1993; Körthauer et al. 1993). In addition to an inability to switch to isotypes dependent on T cells, their antibodies do not undergo affinity maturation, and they do not develop B-cell memory.

Mice lacking CD40L or CD40 have a deficit similar to the human syndrome (Castigli et al. 1994; Kawabe et al. 1994; Renshaw et al. 1994; Xu et al. 1994). In both types of knock-out mice, some TD antigens induce a normal IgM response but no other isotypes, perhaps due to an ability to deliver a signal through sIg. CD40L or CD40 knock-out mice respond perfectly well to TI-2 antigens, e.g., dinitrophenyl (DNP)-Ficoll, making abundant antigen-specific IgM and IgG3, and normal amounts of IgG1, IgG2b and IgA. Sera of non-immunized 3-month-old CD40L knock-out mice contain normal levels of IgM, two- to threefold-reduced levels of IgG2b and IgG3, 10- to 100-fold-reduced levels of IgG1 and IgG2a, and no IgE (Xu et al. 1994). IgE expression in vivo appears to be completely dependent on CD40L.

In vivo, antibody class switching begins about 4 days after activation of naive B cells by TD antigens, initiating in B-cell foci in the T-cell-rich periarteriolar lymphoid sheaths in spleens, and also in B cells present in T-cell zones in lymph nodes (Kraal et al. 1982; Jacob et al. 1991; Weinstein and Cebra 1991; McHeyzer-Williams et al. 1993; Toellner et al. 1996; Toellner et al. 1998). It also occurs, but to a much lower extent, in germinal centers where B cells are also undergoing somatic mutation (Jacob et al. 1991; Liu et al. 1996b). These data suggest that induction of class switching in a TD response occurs during contact with activated $T_H$ cells. IgG, IgA and IgE are made later than IgM during a primary response and account for most of the antibody made during a memory response.

## 2.3 Induction of CSR In Vitro

Cultured mouse sIgM$^+$ B cells can be induced to switch to all isotypes by the B-cell mitogen lipopolysaccharide (LPS), which is considered a TI-1 antigen, in conjunction with certain cytokines (KEARNEY and LAWTON 1975; KEARNEY et al. 1976; ANDERSSON et al. 1978; ISAKSON et al. 1982; SEVERINSON et al. 1982; reviewed by SNAPPER and FINKELMAN 1998 and STAVNEZER 1996a). The mitogen *Staphlococcal aureus* Cowan I (SAC) induces human B cells to switch, but very poorly. Mouse B cells can also be induced to switch in culture with TI-2 antigens, e.g., anti-IgM bound to sepharose beads or soluble anti-IgD dextran (anti-δ dex), if interleukin (IL)-5 is also provided (PURKERSON et al. 1988; SNAPPER et al. 1991; SNAPPER et al. 1992; SNAPPER and MOND 1996). Soluble anti-IgM does not substitute. The role of dextran is not entirely understood. It highly cross-links the anti-δ antibody but also may have a separate stimulatory effect on B cells (BRUNSWICK et al. 1988). TD antigens plus T cells (LEBMAN and COFFMAN 1988; VERCELLI et al. 1989) or CD40L or antibody to CD40 in the absence of antigen also induce switching in cultured mouse and human B cells (JABARA et al. 1990; GASCAN et al. 1991; ROUSSET et al. 1991; DEFRANCE et al. 1992; NOELLE et al. 1992b; SPLAWSKI et al. 1993; FUJIEDA et al. 1995; SNAPPER et al. 1995). The use of antibody to CD40 has allowed the first relatively efficient means for induction of switching in human B-cell cultures.

## 2.4 Isotype Regulation by Cytokines

Cytokines direct and regulate isotype specificity of switching, as demonstrated by analyzing class switching in culture and also in vivo in both normal mice and mice deficient in cytokine genes or their receptors or deficient in molecules involved in cytokine-signal transduction (FINKELMAN et al. 1986; FINKELMAN et al. 1988; KÜHN et al. 1991; HUANG et al. 1993; KAPLAN et al. 1996; SHIMODA et al. 1996; TAKEDA et al. 1996; AKIMOTO et al. 1998).

Table 1 summarizes the effects of cytokines on regulation of CSR to each isotype in mouse B cells activated by different methods, and Table 2 summarizes the regulation of isotype specificity in human B cells activated via CD40. In cultured mouse B cells, LPS alone induces switching to IgG3 and IgG2b (ISAKSON et al. 1982; BERGSTEDT-LINDQVIST et al. 1984; VITETTA et al. 1985; COFFMAN et al. 1986); IL-10 further increases switching to IgG3 (SHPARAGO et al. 1996), and transforming growth factor (TGF)-β1 further increases switching to IgG2b (MCINTYRE et al. 1993). Treatment with LPS + IL-4 induces switching to IgG1 and IgE (BERGSTEDT-LINDQVIST et al. 1984; VITETTA et al. 1985; COFFMAN et al. 1986); LPS + TGF-β1 induces switching to IgA (COFFMAN et al. 1989; SONODA et al. 1989; KIM and KAGNOFF 1990); LPS + interferon-γ (IFN-γ) induces switching to IgG2a (SNAPPER and PAUL 1987). The amounts of switching to each class vary considerably under these conditions, varying from 10% to 40% surface-positive cells for IgG3, IgG2b and IgG1 to less than 2% for IgG2a, IgE and IgA.

**Table 1.** Cytokines that direct class-switch recombination in mouse splenic B-cell cultures stimulated as indicated

| Isotype | LPS | Anti-δ dextran + IL-5 | Anti-δ dextran + LPS + IL-5 | CD40L or antibody to CD40 | Anti-δ dextran + IL-5 + CD40L or anti-CD40 |
|---|---|---|---|---|---|
| IgG3 | IL-4 ↓<br>IFN-γ ↓<br>IL-10 ↑ | γ-IFN ↑ | | | |
| IgG1 | IL-4 ↑<br>γ-IFN ↓ | IL-4 ↑ | | IL-4 ↑ | IL-4 ↑ |
| IgG2b | TGF-β1 ↑<br>IL-4 ↓ | | | | |
| IgG2a | IFN-γ ↑ | IFN-γ ↑ | | | |
| IgE | IL-4 ↑<br>IFN-γ ↓<br>TGF-β1 ↓ | No switching to IgE | IL-4 ↑ | IL-4 ↑ | IL-4 ↑ |
| IgA | TGF-β1 ↑<br>+IL-4 ↑↑ | TGF-β1 ↑<br>+IL-4 ↑↑ | TGF-β ↑<br>+IL-4 ↑↑<br>IL-10 ↓ | | TGF-β1 ↑<br>+IL-4 ↑↑ |

IFN, interferon; Ig, immunoglobulin; IL, interleukin; LPS, lipopolysaccharide; TGF, transforming growth factor

**Table 2.** Cytokines that direct class-switch recombination in human B cells

| Isotypes | T cells or anti-CD40 |
|---|---|
| IgG1 | IL-4 ↑<br>IL-10 ↑ |
| IgG3 | IL-4 ↑<br>IL-10 ↑ |
| IgG2 | |
| IgG4 | IL-4 ↑<br>IL-13 ↑ |
| IgE | IL-4 ↑<br>IL-13 ↑<br>TNFα ↑<br>TGF-β1 ↓ |
| IgA1 | TGF-β1 ↑<br>IL-10 ↑ |
| IgA2 | TGF-β1 ↑<br>IL-10 ↑ |

Tonsillar or peripheral blood B cells stimulated as indicated.
IL, interleukin; TGF, transforming growth factor; TNF, tumor necrosis factor

In the presence of the appropriate cytokines, class switching to all isotypes, except IgE, can also be induced by mimics of TI-2 antigens, e.g., anti-δ dex plus IL-5 (Snapper and Finkelman 1998). IL-5 is essential for induction of switching to all isotypes when anti-IgM coupled to Sepharose beads or anti-δ dex are used to stimulate switching, and appears to actually induce recombination (Mandler et al. 1993a; Snapper and Mond 1993). Some switching to IgG1 is obtained in B cells cultured with anti-δ dex + IL-5, and IL-4 further increases switching to IgG1. When switching is induced by anti-δ dex + IL-5, IFN-γ is required for switching to IgG2a, but also to IgG3 (SNAPPER et al. 1992), and TGF-β1 is required for IgA switching. Induction of switching to IgA is induced best by a combination of LPS + anti-δ dex + IL-5 along with TGF-β1 (MCINTYRE et al. 1995). IL-4 addition further increases IgA switching.

IL-4 + CD40L (soluble or expressed on insect cells) or antibody to CD40 induce switching to IgG1 and to IgE in mouse-splenic B cells (LANE et al. 1993; SNAPPER et al. 1995) and to all IgG subclasses (except IgG2) and to IgE in human B cells (JABARA et al. 1990; GASCAN et al. 1991; SPLAWSKI et al. 1993; FUJIEDA et al. 1995). Although, in these in vitro experiments, signals delivered via sIg were not found necessary, addition of anti-δ dex + IL-5 further increases switching induced by CD40L + IL-4 to IgG1 (SNAPPER et al. 1995). CD40L can replace LPS for induction of equivalent optimal switching to IgA in cultures also containing anti-δ dex + IL-5 + TGF-β1 + IL-4 (MCINTYRE et al. 1995).

# 3 Regulation of Germline (GL) Transcripts Directs CSR

## 3.1 Regulation of Germline Transcripts by Cytokines

The cytokines IL-4, IL-10, IFN-γ and TGF-β1 regulate CSR by regulation of transcripts from the unrearranged $C_H$ genes prior to class switching, yielding what are called germline (GL) transcripts (STAVNEZER-NORDGREN and SIRLIN 1986; YANCOPOLOUS et al. 1986; LUTZKER et al. 1988; STAVNEZER et al. 1988; SIDERAS et al. 1989; SEVERINSON et al. 1990; COLLINS and DUNNICK 1993; JUNG et al. 1993; ZHANG et al. 1993; FUJIEDA et al. 1996). When examined, this regulation has been found to occur at the transcriptional level (discussed below). IL-4 and IL-10 also regulate CSR at additional levels. Induction or suppression of GL transcription by particular cytokines directly correlates with subsequent switching to the same isotype after addition of a B-cell activator, e.g., LPS, CD40L or anti-δ dex (STAVNEZER et al. 1988; ROTHMAN et al. 1988; BERTON et al. 1989; ESSER and RADBRUCH 1989; GAUCHAT et al. 1990; LEBMAN et al. 1990; SEVERINSON et al. 1990; ISLAM et al. 1991; SHOCKETT and STAVNEZER 1991; WEINSTEIN and CEBRA 1991; BERTON and VITETTA 1992; GAFF et al. 1992; GOODMAN et al. 1993; TURAGA et al. 1993; WAKATSUKI and STROBER 1993; KITANI and STROBER 1994; XU and ROTHMAN 1994; FUJIEDA et al. 1996).

All the GL transcripts have an analogous structure, with initiation sites located upstream of each switch region at the so-called I exons, and all are transcribed through the switch region. The transcripts are polyadenylated and spliced, with the I exon being spliced to the normal $C_H$ acceptor. Figure 1 shows a diagram of the Ig heavy-chain genes in cells expressing IgM and IgD, the structure of the GL ε transcript and CSR to IgE. Each of the unrearranged $C_H$ genes is transcribed in activated B cells treated with the same cytokine that direct switching to that $C_H$ gene. GL transcripts direct CSR but are not sufficient to induce it.

## 3.2 Accessibility Model

The correlation of expression of GL transcripts with CSR to the same $C_H$ gene led to the proposal that accessibility of chromatin regulates CSR. The hypothesis is that CSR is directed to specific switch regions that have been induced to become accessible by *trans*-acting factors (Stavnezer et al. 1984; Stavnezer-Nordgren and Sirlin 1986; Yancopolous et al. 1986). The $C_H$ genes to which switching occurs have been shown to be hypomethylated and to have DNase-hypersensitive sites within the promoter for the GL transcripts (Stavnezer-Nordgren and Sirlin 1986; Schmitz and Radbruch 1989; Berton and Vitetta 1990). In addition, DNase-hypersensitive sites are detected in the Sγ1 regions of splenic B cells induced for 2 days to switch with IL-4 and CD40L (Cunningham et al. 1998). The accessibility model is in agreement with the finding that switch recombination frequently occurs on both chromosomes and to the same isotype in both mice and humans, consistent with regulation by *trans*-acting factors (Hummel et al. 1987; Winter et al. 1987; Kepron et al. 1989; Irsch et al. 1993). The accessibility model is a very general model, and recently several types of experiments discussed below have refined the model, but it is still unclear what constitutes accessibility for CSR. The data discussed below suggest that accessibility to transcription factors, which is the accepted meaning of accessibility in other contexts, is not sufficient to enable a particular switch region to be chosen for CSR.

## 3.3 Germline Transcripts are Required for Switching

The fact that all GL transcripts have the same overall structure suggests that the transcripts themselves might have a function. Targeted mutations that ablate or alter the structure of GL transcripts further indicate that the GL transcripts themselves are required for switching (Jung et al. 1993; Zhang et al. 1993; Bottaro et al. 1994; Lorenz et al. 1995). Deletion or replacement of the promoter and I exons of unrearranged γ2b or γ1 genes with a neomycin-resistance ($neo^r$) gene transcribed in the antisense direction results in no switching on that chromosome to that particular $C_H$ gene, although the other allele and other $C_H$ genes on that chromosome are unaffected (Fig. 2; Jung et al. 1993; Zhang et al. 1993). Sur-

→

**Fig. 2.** Targeted mutations in the unrearranged immunoglobulin $C_H$ genes and their effect on class switching. Diagrams of the targeted mutations that have been created in the unrearranged Ig heavy-chain genes and their effects on class switching. See text for further description

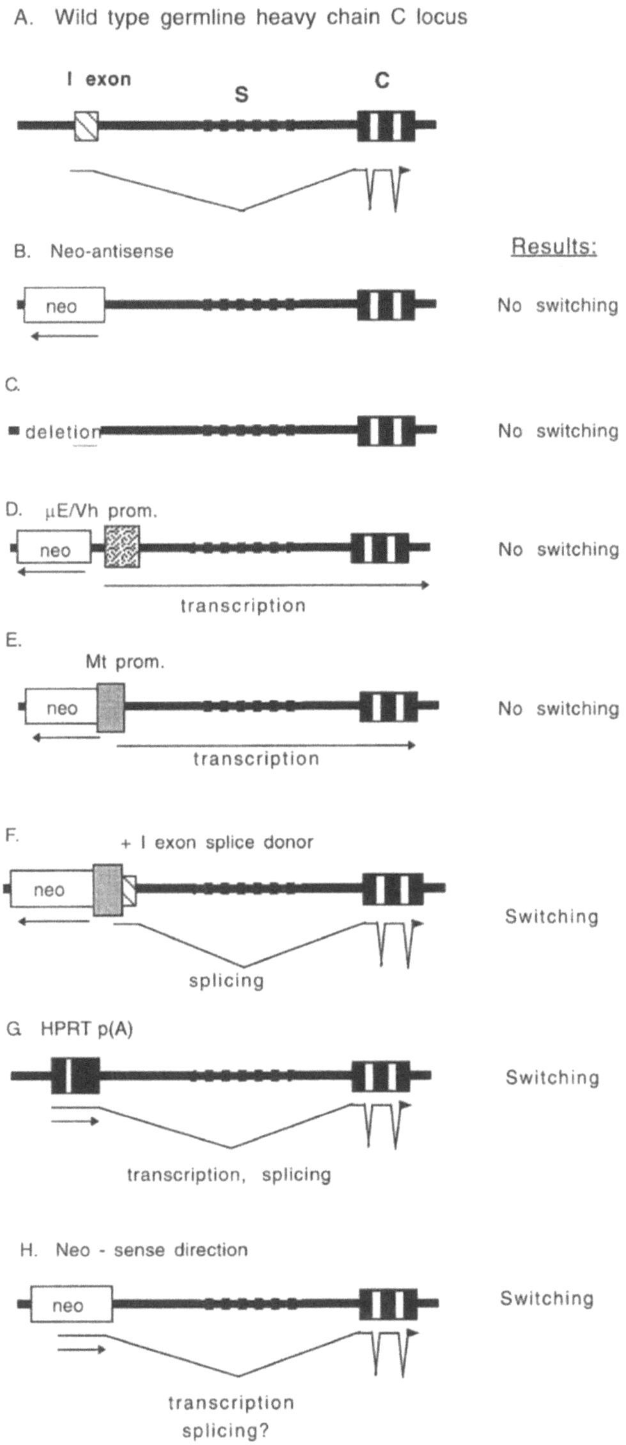
A. Wild type germline heavy chain C locus
I exon
S
C
B. Neo-antisense
Results:
neo
No switching
C.
deletion
No switching
D. μE/Vh prom.
neo
No switching
transcription
E.
Mt prom.
neo
No switching
transcription
F.
+ I exon splice donor
neo
Switching
splicing
G. HPRT p(A)
Switching
transcription, splicing
H. Neo - sense direction
neo
Switching
transcription
splicing?

prisingly, when the Iε or Iγ1 segments are replaced by heterologous strong promoters (metallothionein or Ig $V_H$, respectively), driving transcription in the normal sense direction, only very low levels of switching occur to that $C_H$ gene (Fig. 2; BOTTARO et al. 1994; LORENZ et al. 1995). These data indicate that transcription per se is not sufficient to direct switching to the switch region.

However, if the 114-bp DNA segment containing the splice donor for the Iγ1 exon is inserted immediately 3′ to the heterologous metallothionein promoter in its original position upstream of Sγ1, switching is recovered (Fig. 2; LORENZ et al. 1995). LPS alone induces switching to IgG1 at the targeted allele. Insertion of the splice donor greatly increases the rate of transcription driven by the metallothionein promoter, as measured in nuclear run-on assays, and also greatly increases the steady-state levels of the RNA produced as assayed by Northern blotting. The increased transcription rate and stability of GL RNA do not appear to account for the increased ability to switch, however, since in the experiments of BOTTARO et al. (1994; Fig. 2), in which the Ig-μ intron and $V_H$ promoter is inserted into the Iε locus in the absence of a splice donor, abundant amounts of large, probably unspliced transcripts are detected and yet switching does not occur. These data suggest that splicing of the transcript or the transcript itself is required for CSR. However, it was possible that the 114-bp Iγ1 segment containing the splice donor provided a binding site for a protein complex required for CSR.

Recent results obtained from two other knock-out mice suggest that the specific sequence of the I exon is not important. These data further support the requirement for a splice donor in the GL transcript, although the function of the splice donor in CSR is still completely unclear. In mice created by HARRIMAN et al. (1996), the GL Iα exon is replaced by a mini hypoxanthyl ribosyl transferase (HPRT) gene in the sense direction (Fig. 2). Replacement of the Iα exon with the HPRT gene does not reduce switching to IgA, as these mice have normal levels of serum IgA, and B cells from these mice are induced by LPS to switch to IgA. TGF-β1 does not further increase switching. The HPRT-gene segment includes a polyadenylation site, and B cells from these mice express abundant HPRT mRNA. They also express two RNA species derived by splicing of the HPRT exons to the normal splice acceptor for Cα (QIU et al. 1999). The amount of these spliced transcripts is about three times greater than the normal GL α transcript present in wild-type B cells and, thus, their levels are probably sufficient to support switching to IgA by these B cells. In addition, SEIDL et al. (1998) have shown that switching to IgG2b occurs on the targeted allele, in which a $neo^r$ gene in the sense orientation replaces the Iγ2b exon (Fig. 2). Transcription through the Sγ2b region occurs in these B cells, although it is unknown if the transcript is spliced.

Altogether, these data suggest that the particular sequence of the I exon is not important, although the overall structure of the transcript may be essential. It is possible that splicing of the I and $C_H$ exons may be required for CSR, perhaps because splicing factors are involved or perhaps because splicing cuts the transcript, releasing the intron to associate with the DNA of the switch region (COLLIER et al. 1988; STAVNEZER 1996a).

## 3.4 Cytokines Regulate GL RNA at the Transcriptional Level

Nuclear run-on assays and RNA stability measurements indicate that regulation of expression of GL transcripts mostly takes place at the transcriptional level (ROTHMAN et al. 1991; SHOCKETT and STAVNEZER 1991; LORENZ et al. 1995). Furthermore, transient or stable transfection of reporter genes driven by 5′ flanking segments of GL mouse Iγ1 and both mouse and human Iε and Iα exons demonstrate that transcription driven by each of these promoters is inducible by the cytokine which directs switching to that antibody class (ROTHMAN et al. 1991; LIN and STAVNEZER 1992; XU and STAVNEZER 1992; ICHIKI et al. 1993; NILSSON and SIDERAS 1993; ALBRECHT et al. 1994; DELPHIN and STAVNEZER 1995; WARREN and BERTON 1995).

### 3.4.1 Interleukin-4

Consistent with its effect on CSR, IL-4 induces GL γ1 and ε transcripts and inhibits GL γ2b and γ3 transcripts in activated mouse B cells (LUTZKER et al. 1988; STAVNEZER et al. 1988; BERTON et al. 1989; ESSER and RADBRUCH 1989; LUNDGREN et al. 1989; ROTHMAN et al. 1990; SEVERINSON et al. 1990; XU and ROTHMAN 1994; WARREN and BERTON 1995). IL-4 induces γ4 and ε GL transcripts in activated human B cells (GAUCHAT et al. 1990; SHAPIRA et al. 1992; KITANI and STROBER 1993). A mouse lacking the IL-4 gene expresses greatly reduced levels of IgE in response to TD antigens, although expression of IgG1 is only reduced two-fold, probably because GL γ transcripts can be induced by other means (see below; KÜHN et al. 1991). Interestingly, IgG2b, IgG2a and IgG3 are all increased about fivefold in TD responses in these mice. This is probably due to the increase in the Th1 type of $T_H$ cells and the resultant increase in IFN-γ in these mice (KOPF et al. 1993).

The mechanism of induction of transcription by IL-4 of the GL ε and γ1 transcripts in mouse and human B cells has been shown to be due to activation and binding of Stat6 to these promoters, as is true for other IL-4 responsive promoters (KOHLER and RIEBER 1993; HOU et al. 1994; LUNDGREN et al. 1994; SCHINDLER et al. 1994; DELPHIN and STAVNEZER 1995; FENGHAO et al. 1995; WARREN and BERTON 1995; MIKITA et al. 1996; REICHEL et al. 1997; MIKITA et al. 1998; SHEN and STAVNEZER 1998). Stat6-deficient mice do not express IgE (KAPLAN et al. 1996; SHIMODA et al. 1996; TAKEDA et al. 1996). As is true for all Stat proteins, Stat6 becomes phosphorylated by the associated Jak tyrosine kinases upon engagement of its associated cytokine receptor, dimerizes and moves very rapidly into the nucleus, where it binds DNA (DARNELL 1997; O'SHEA 1997). Stat6 is activated by Jak1 and Jak3, which are associated with the IL-4 receptor. Like other IL-4-responsive promoters, the IL-4-responsive elements in both the γ1 and ε promoters contain a binding site for Stat6 (previously called IL-4 Stat; LUNDGREN et al. 1994; BERTON and LINEHAN 1995; DELPHIN and STAVNEZER 1995; FENGHAO et al. 1995).

Unlike other Stat proteins, Stat6 cannot activate transcription by itself, but (in the cases thus far examined) requires an adjacent binding site for CCAAT/

enhancer-binding protein (C/EBP), a member of the large basic "zipper" (b-Zip) transcription factor family, closely related to the AP-1 family (Xu and Stavnezer 1992; Lundgren et al. 1994; Delphin and Stavnezer 1995; Mikita et al. 1996; Mikita et al. 1998). The C/EBP binding site has been shown to be essential for expression of the GL mouse γ1 and mouse and human ε promoters and for their induction by IL-4. Although transfected over-expressed C/EBPβ has been shown to bind and activate the human GL ε promoter, it is not clear which protein(s) binds this site in vivo and is required for transcription of the promoters. Mature mouse $sIg^+$ B cells have little or no C/EBPβ (Cooper et al. 1994). Furthermore, C/EBPβ-deficient mice produce abundant IgG1 (Screpanti et al. 1995).

Expression and IL-4 induction of the GL ε and γ1 promoters also requires the presence of two (ε promoter) or three (γ1 promoter) binding sites for nuclear factor (NF)-κB, located within 80 nucleotides 3′ to the Stat6 site (Xu and Stavnezer 1992; Delphin and Stavnezer 1995; Lin and Stavnezer 1996; Iciek et al. 1997; Messner et al. 1997). Although NF-κB is not activated by IL-4 to enter the nucleus and to bind DNA, it binds cooperatively with Stat6 to the GL ε promoter containing both Stat6 and κB sites, and synergizes with Stat6 to induce transcription in IL-4-treated B cells (Iciek et al. 1997; Shen and Stavnezer 1998). These data help explain why B cells cannot transcribe GL ε transcripts and switch to IgE unless they contain activated NF-κB. One method of activating NF-κB is signaling via CD40, as occurs in TD responses. Figure 3 shows the nucleotide sequences and identified transcription-factor binding sites in the IL-4/CD40-responsive regions of the mouse GL γ1 and ε promoters.

The Stat6-binding sites that have been most thoroughly studied, those in the mouse GL γ1 and ε promoters, reside about 120–150 nucleotides upstream of the first RNA initiation sites. In addition to this upstream Stat6 site, two non-consensus sites, one near the RNA initiation site of the GL ε promoter and one located 3′ to the Iγ1 exon, have been shown to weakly bind Stat6 in electrophoretic-mobility-shift assays (EMSAs; Schindler et al. 1994; Cunningham et al. 1998). Whether these sites are involved in regulating expression of the GL transcripts has not been reported.

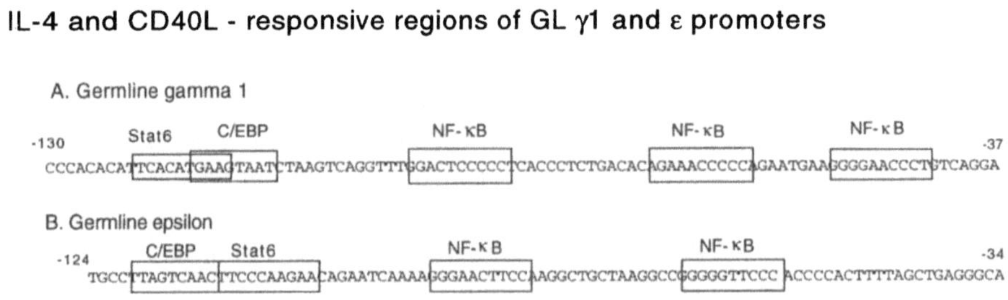


**Fig. 3.** Comparison of IL-4/CD40 responsive regions of mouse GL γ1 and ε promoters

### 3.4.2 Interferon-γ

Addition of IFN-γ to LPS-activated B cells induces GL γ2a transcripts (SEVERINSON et al. 1990; COLLINS and DUNNICK 1993). The effects, however, of IFN-γ on GL γ2a transcripts and switching to IgG2a are not as dramatic as the effects of IL-4 on switching to IgG1 or IgE. IFN-γ also induces GL γ3 transcripts and switching to IgG3 in B cells activated with anti-IgD dextran + IL-5 (ZELAZOWSKI et al. 1995). IFN-γ inhibits induction by LPS + IL-4 of GL γ1 and ε transcripts and switching to IgG1 and IgE by mouse-splenic B cells (SNAPPER and PAUL 1987; BERTON et al. 1989; XU and ROTHMAN 1994). The inhibition of GL γ1 and ε transcripts by IFN-γ is only partial, however. IFN-γ has little effect on induction of IgE production by human B cells by anti-CD40 antibody and IL-4 (GASCAN et al. 1991). A mouse lacking the receptor for IFN-γ has reduced antigen-specific IgG2a (fivefold) and IgG3 (twofold) responses to vaccinia virus (HUANG et al. 1993). Since the inhibition is incomplete, there must be additional mechanisms for inducing γ2a GL transcripts and switching to IgG2a.

The mechanisms for induction of transcription of GL γ3 and γ2a transcripts and inhibition of IL-4 induced γ1 and ε transcripts by IFN-γ are unknown. IFN-γ is known to activate transcription by activation of Stat1 which binds to gamma-activated sequence elements in DNA (TTCN$_3$GAA), but binds only poorly to Stat6-binding sites (TTCN$_4$GAA) due to the extra nucleotide between the palindromic sequences (SCHINDLER et al. 1995). Functional Stat1-binding sites have not been identified in any of the GL-transcript promoters, although the weak Stat6-binding site near the RNA initiation site of the GL ε promoter binds Stat1 with apparent high affinity (SCHINDLER et al. 1994). Furthermore, IFN-γ was found to only marginally inhibit a reporter plasmid driven by the mouse GL ε promoter (Delphin and Stavnezer, unpublished observations) and, thus, it is possible that sequences outside of the proximal 5′ flanking sequences studied in the ε reporter plasmids may be required for inhibition by IFN-γ.

There also may be additional levels of inhibition of IgE expression by IFN-γ. This possibility is suggested by the fact that GL ε transcripts were detected in IL-4-deficient mice infected for 11 days with *Nippostrongylus brasiliensis*, although productive ε mRNA was not detected unless the mice were also injected with anti-IFN-γ (MORAWETZ et al. 1996). This result suggests that inhibition of IgE switching by IFN-γ may involve a process other than regulation of GL ε transcripts. Alternatively, the levels of GL ε transcripts were increased by the anti-IFN-γ antibody to levels sufficient for class switching.

### 3.4.3 Transforming Growth Factor-β1

Consistent with the requirement for TGF-β1 for switching to IgA in splenic B cells induced by LPS, TGF-β1 induces GL α transcripts in mouse and human B cells and B cell lines treated with LPS or CD40L (LEBMAN et al. 1990; ISLAM et al. 1991; SHOCKETT and STAVNEZER 1991; STAVNEZER 1995a). TGF-β1 also increases GL γ2b transcripts and switching to IgG2b by LPS-activated B cells, but since quite low levels of TGF-β1 are required for IgG2b switching, and since activated B cells synthesize active TGF-β1, addition of exogenous TGF-β1 does not always increase the level of switching to IgG2b (McIntyre et al. 1993). Whether exogenous TGF-β1 increases switching to IgG2b depends on the strain of mouse. TGF-β1 also specifically inhibits switching to IgE, because it inhibits induction of GL ε transcripts by IL-4 (Shockett and Stavnezer 1991; Gauchat et al. 1992).

Splenic B cells treated with TGF-β1 + LPS do not switch very well to IgA, although if one also adds anti-δ dex, IL-5 and IL-4, about 10% of splenic B cells switch to IgA within 4 days (McIntyre et al. 1995). CD40L can be substituted for LPS in these cultures. The requirement for some of these activators may partly be due to the inhibitory effect of TGF-β1 on proliferation (reviewed in Stavnezer 1995b), although the efficacy of these additional reagents is not solely due to their ability to stimulate DNA synthesis (see below).

TGF-β1 has been demonstrated to induce GL α transcripts at the transcriptional level in the mouse I.29μ B-cell lymphoma, which switches to IgA when treated with LPS + TGF-β1 (Shockett and Stavnezer 1991). TGF-β1 also induces reporter plasmids driven by the mouse and human GL α promoters (Lin and Stavnezer 1992; Nilsson and Sideras 1993). The nucleotide sequences of the human α1 and α2 and mouse α promoters are highly similar, and these promoters appear to be regulated similarly. A novel TGF-β1-responsive element (Tβ-RE) is present in all three promoters (Lin and Stavnezer 1992). The Tβ-RE consists of a direct repeat located at approximately −110 bp relative to the first RNA initiation site for all three promoters. The Tβ-RE, along with a putative DNA binding site for ATF and a demonstrated binding site for Ets-family proteins, together appear to be the most important binding sites for regulating the GL α promoters in reporter-gene assays.

It has been difficult to determine which proteins transduce the TGF-β1 signal to the GL α promoter. The Tβ-RE has recently been found to contain two binding sites for core binding factor (CBF; also known as acute-myelogenous-leukemia protein or polyoma-enhancer-binding protein) and two binding sites for Smad proteins (Shi and Stavnezer 1998; Xie et al. 1999; Sideras, personal communication; Ito, personal communication). TGF-β1 treatment of I.29μ cells and splenic B cells induces synthesis of one member of the CBFα family (CBFα3) and increases the binding of CBF to the mouse GL α promoter (Shi and Stavnezer 1998). Binding of Smads to the promoter has been demonstrated using recombinant Smad proteins (Ito, personal communication). It is very difficult to detect binding of Smads to DNA, not just to the GL α promoter, using nuclear extracts. There is at least one additional CBF-binding site and one additional putative Smad-binding site located 3′ to this repeat element in the mouse α promoter (Shi and Stavnezer 1998; Shi, unpublished).

Smads are a recently identified family of proteins, which are rapidly activated by ser–thr phosphorylation after ligand engagement of the TGF-β1-receptor family (reviewed by Heldin et al. 1997 and Massague et al. 1997). Smad-family proteins form complexes which bind DNA, although the binding sites are only beginning to be defined (Zawel et al. 1998).

CBF proteins, however, were not previously known to be induced by TGF-β1, although they have been shown to be involved in tissue-specific differentiation in the hematopoietic lineage (reviewed in Meyers and Hiebert 1995 and in Speck and Stacy 1995). Although the studies are incomplete, there appear to be interactions between several of the proteins which bind the GL α promoter, as mutations of several binding sites interfere with induction by TGF-β1 (Shi and Stavnezer, unpublished).

### 3.4.4 Interleukin-10

IL-10 increases GL γ1 and γ3 transcripts and switching to IgG1 and IgG3 in human $IgD^+$ B cells activated by antibody to CD40 (Fujieda et al. 1996; Malisan et al. 1996). IL-10 also has effects on switching in mouse B cells, but they differ from the effects in human cells. Although IL-10 stimulates switching to IgG3 in mouse B cells (shown by both surface IgG3 expression and by the digestion-circularization polymerase chain reaction (DC-PCR) assay for switch recombination (Chu et al. 1992), it has no effect on GL γ3 transcripts (Shparago et al. 1996). It also inhibits switching to IgA in mouse B cells stimulated with LPS + TGF-β1 + IL-4 + anti-δ dex + IL-5, again without affecting levels of GL-α transcripts. A hint of a possible mechanism for its effects in mouse B cells is that it inhibits DNA synthesis ($^3$H-thymidine incorporation) in B cells treated with LPS under conditions for inducing IgG3 or IgA switching. Supporting this hypothesis is the finding that IL-10 does not inhibit IgA switching when CD40L is substituted for LPS, nor does it inhibit DNA synthesis under these conditions.

### 3.4.5 Interleukin-13

IL-13 transmits a signal in human B cells through a receptor which includes the IL-4R α chain (Callard et al. 1997). Thus, IL-13 mimics many of the functions of IL-4 in human B cells, including an ability to induce GL ε transcripts and to direct switching to IgG4 and IgE in B cells induced by membranes from activated $T_H$ cells (Punnonen et al. 1993). However, IL-13 does not appear to induce IgE production in mouse B cells (Zurawski and Vries 1994; Morawetz et al. 1996).

### 3.4.6 Tumor Necrosis Factor α

Tumor necrosis factor (TNF) α synergizes with IL-4 to induce GL ε transcripts and IgE production by human B cells (Gauchat et al. 1992). The mechanism of this effect may be by the activation of NF-κB. TNFα is known to activate NF-κB (reviewed by Ghosh et al. 1998), and NF-κB activation induces GL ε transcripts and synergizes with IL-4 to further induce GL ε transcripts (Iciek et al. 1997; Messner et al. 1997; Shen and Stavnezer 1998). The mechanism of induction of GL-ε transcripts by NF-κB is discussed in the next section.

## 3.5 CD40 Signaling Also Induces Transcripts of Unrearranged $C_H$ Genes

Membranes from activated mouse $T_H$ cells can induce GL γ1 transcripts in the absence of IL-4 (i.e., using membranes from T cells of IL-4 knock-out mice), although addition of IL-4 further increases the level of the transcripts (SCHULTZ et al. 1992). Membranes from activated $T_H$ cells did not induce GL γ3, γ2b, ε or α transcripts, although they could synergize with IL-4 to induce GL ε transcripts. Furthermore, CD40 signaling itself has been shown to induce GL γ1 and ε transcripts in mouse-splenic B cells (WARREN and BERTON 1995; FERLIN et al. 1996; LIN et al. 1998; LINEHAN et al. 1998) and induction of transcription from the GL γ1 and ε promoters in reporter-gene constructs (WARREN and BERTON 1995; LIN and STAVNEZER 1996; ICIEK et al. 1997; LIN et al. 1998). Since CD40 signaling induces some, but not all GL transcripts, it contributes to specificity of isotype switching.

The finding that CD40 signaling induces GL γ1 and ε transcripts in the absence of IL-4 provides one possible explanation for how IgG1 and IgE expression can be induced in response to TD antigens and infection by parasites in IL-4-deficient mice (KÜHN et al. 1991; VONDERWEID et al. 1994; HJULSTROM et al. 1995; MORAWETZ et al. 1996; NOBEN-TRAUTH et al. 1996). In human systems the data are controversial, but in general it is found that anti-CD40 antibody alone does not induce GL transcripts in human B cells, although it synergizes with IL-4 to induce ε and all the γ transcripts (except γ2) and switching to the corresponding isotypes (GAUCHAT et al. 1992; JUMPER et al. 1994; KITANI and STROBER 1994; FUJIEDA et al. 1995).

The mechanism of induction of GL ε and γ1 transcripts in mouse B cells by CD40 signaling has been shown to be the activation of NF-κB to bind to the two and three κB sites in the γ1 and ε promoters, respectively (LIN and STAVNEZER 1996; ICIEK et al. 1997; LIN et al. 1998). CD40 signaling is known to induce three families of transcription factors, AP-1, NF-κB and NF-AT (LALMANACH-GIRARD et al. 1993; BERBERICH et al. 1994; CHOI et al. 1994; FRANCIS et al. 1995; HUO and ROTHSTEIN 1995). However, it was found that NF-κB, but neither AP-1 nor NF-AT, binds to the CD40-responsive regions of the GL γ1 and ε promoters. NF-κB/Rel binding is required for any activity of the GL ε promoter and is also important for expression of the γ1 promoter, although it is not essential for some expression of the GLγ1 transcripts themselves (SNAPPER et al. 1996b; LIN et al. 1998).

The binding of NF-κB to the two κB sites in the GL ε promoter is required for IL-4 induction of the promoter, although IL-4 does not activate NF-κB (DELPHIN and STAVNEZER 1995; ICIEK et al. 1997; SHEN and STAVNEZER 1998). Stat6 and NF-κB bind cooperatively to the ε promoter and synergistically activate transcription from the ε promoter in reporter-gene assays, thus providing a mechanistic explanation for how IL-4 and CD40L synergistically induce GL-ε RNA (WARREN and BERTON 1995; ICIEK et al. 1997; MESSNER et al. 1997; LINEHAN et al. 1998; SHEN and STAVNEZER 1998). Synergy between Stat6 and NF-κB for induction of the GL γ1 promoter has not been examined.

Unlike CD40L, LPS given alone does not induce GL γ1 transcripts in mouse-splenic B cells, although LPS also activates NF-κB. This is explained by the findings that CD40L and LPS activate different members of the NF-κB/Rel family and that the different members of the family have different abilities to induce transcription from the GL γ1 promoter (LIN et al. 1998). CD40L treatment causes a sustained (> 24h) induction of RelA (p65), RelB, c-Rel and p50, whereas LPS treatment results in a transient activation of RelA, RelB and p50 but a sustained activation of c-Rel. Unlike RelA-p50 and RelB-p50 heterodimers, c-Rel-p50 heterodimers are unable to activate the GL γ1 promoter. Furthermore, the data indicate that the c-Rel-p50 heterodimers compete with the RelA- and RelB-p50 heterodimers for binding to the GL γ1 promoter. Thus, the activity of the promoter is regulated by the amount of the transactivating heterodimers (RelA-p50 and RelB-p50) and by their ratio to the inhibitory heterodimer (c-Rel-p50) (LIN et al. 1998). Unlike the γ1 promoter, the mouse GL-ε promoter is activated by c-Rel-p50 heterodimers, although not as well as by RelA-p50 heterodimers. These data help explain why TD but not TI-1 antigens selectively induce IgG1 responses.

Consistent with the finding that NF-κB/Rel proteins bind and activate the GL γ1 and ε promoters, mice deficient in various members of the NF-κB family have been found to have defects in CSR. NF-κB/p50 knock-out mice have impaired ability to express some antibody classes in sera of unimmunized mice and in antigen-specific antibodies secreted in response to a TD antigen (SHA et al. 1995). In the TD response examined, IgE was not produced, IgA was reduced by 4-fold, IgG1 and IgG2b by 30-fold and IgG3 by 6-fold. Evidence discussed above indicates that the reduction in IgE and IgG1 is due, at least in part, to the requirement for NF-κB/Rel heterodimers for expression of GL-ε transcripts and for maximal levels of GL-γ1 transcripts. Furthermore, a consensus NF-κB-binding site has been identified in the GL-γ3 promoter (GERONDAKIS et al. 1991).

SNAPPER et al. (1996b) demonstrated that cultured B cells from p50-deficient mice are not activated to express GL-γ3 or -ε transcripts or to switch to IgG3 or IgE under conditions that in wild-type mice induce switching to IgG3 (anti-δ dex + IL-5 + IFN-γ) or to IgE (CD40L + IL-4 + IL-5). However, they did observe GL γ1 transcripts and switching to IgG1. It appears likely that IL-4 induces the GL γ1 transcripts in the p50-deficient B cells since IL-4 induction of GL γ1 transcripts does not require NF-κB (LIN et al. 1998). Also, anti-δ dex + IL-4 + IL-5 induces switching to IgG1 in these B cells. B cells from these mice were shown to synthesize DNA in response to anti-IgM, but not in response to LPS (SHA et al. 1995). Apparently, induction of CSR by CD40L or by anti-δ dex and IL-5 does not require NF-κB p50. The reason that IgG1 may be so greatly reduced in vivo in TD responses is probably that in vivo T cell help is delivered locally to antigen activated B cells (TOELLNER et al. 1998) and, thus, IL-4 may be limiting in vivo. Switching to IgA in response to a TD antigen is reduced in the p50-deficient mice and nearly eliminated in splenic B cells activated under optimal conditions for IgA switching. GL α transcripts are, however, expressed at levels comparable to wild-type mice, consistent with the finding that the α promoter is not activated by NF-κB (SNAPPER et al. 1996b; Shi and Stavnezer, unpublished observations). These data demonstrate that NF-κB/Rel is involved in regulating CSR to IgA at other levels than those involved in induction of GL α transcripts.

Mice deficient in RelB show fourfold reduced IgG1 levels in response to a TD antigen, and greatly reduced IgE levels in unimmunized mice (WEIH et al. 1997). Other isotypes were not examined in this report. When B cells from these mice are activated to switch in culture under optimal conditions, which include IL-4 for IgG1, switching to all isotypes is nearly normal, reduced at most by twofold (SNAPPER et al. 1996a). The difference in the effects of the p50 and Rel B knock-out mutations is presumably due to the fact that RelA and/or c-Rel can substitute for RelB, whereas p50 is required for most NF-κB activity.

Disruption of the *relA* gene causes embryonic lethality (BEG et al. 1995). severe combined immunodeficient (SCID) mice reconstituted with $RelA^{-/-}$ fetal liver cells have a 10-fold decrease in serum IgG1, a 100-fold decrease in serum IgA, but no effect on any other isotype compared with SCID mice receiving wild-type fetal-liver cells (DOI et al. 1997). The responses to immunization were not examined. The finding that IgG1 and IgA are reduced agrees with results for other NF-κB proteins. The finding that IgE production does not require RelA suggests that other NF-κB/Rel proteins are sufficient for activation of the GL ε promoter.

B cells from mice bearing a mutation in which the transactivation domain of c-Rel is deleted, but the DNA-binding domain is retained, do not express GL γ1 transcripts in response to LPS and IL-4 (ZELAZOWSKI et al. 1997a). These results appear to contradict results described above which indicate that c-Rel is not required for transcription of GL γ1 RNA. However, it is possible that the c-Rel DNA-binding domain in these B cells competes with the transactivating RelA-p50 and RelB-p50 dimers for binding to the GL γ1 promoter and thereby inhibits the promoter. Thus, the phenotype of these mice may not indicate the specific effects of c-Rel but rather the effects of all NF-κB proteins. These B cells do not express GL γ3 transcripts and do not switch to IgG3, consistent with the results in p50-deficient mice. Also consistent with data from the p50-deficient mice, these B cells synthesize GL α transcripts when induced with LPS + TGF-β1 but do not switch to IgA. Interestingly, the c-Rel-mutant B cells can be induced with IL-4 + LPS to express GL ε transcripts, suggesting that the c-Rel DNA-binding domain cannot inhibit the binding of the other NF-κB/Rel proteins to the GL ε promoter. However, these B cells do not switch to IgE, indicating that NF-κB/Rel proteins are required at an additional level for CSR to IgE, similar to the results for IgA switching. A different line of mice, entirely lacking c-Rel, have major defects in T cell function and cytokine production. Ig isotype production has not been thoroughly examined in these mice (KÖNTGEN et al. 1995).

## 3.6 Cross-Linking sIg May Regulate Isotype Specificity

In the absence of IL-5, anti-δ dex, a TI-2 mimic, induces GL γ1, γ3 and γ2b transcripts, but not γ2a or ε transcripts in mouse-splenic B cells (ZELAZOWSKI et al. 1995). Furthermore, anti-δ dex specifically increases CSR to IgA and IgG1 when added to cultures containing LPS (or CD40) + IL-4 + IL-5 (MCINTYRE et al. 1995) and increases the levels of GL γ1 and α transcripts (SNAPPER and FINKELMAN 1998).

Thus, like CD40 signaling, the signal from anti-δ dex appears to contribute to isotype specificity of class switching. Antibody to IgD that is not conjugated to dextran will not substitute to induce GL transcripts or CSR.

The finding that anti-δ dex induces GL γ1 transcripts suggests that cross-linking sIg by antigen may provide signals, in addition to CD40 signals, that contribute to the nearly normal levels of IgG1 switching in IL-4 deficient mice (Kühn et al. 1991; Morawetz et al. 1996). This is also supported by the finding that reporter plasmids driven by the GL γ1 promoter are activated by phorbol ester, a mimic of part of the signal delivered through sIg (Xu and Stavnezer 1992).

## 3.7 Pax-5/B-Cell-Specific Activator Protein (BSAP)

Pax-5/B-Cell-Specific Activator Protein (BSAP) is a transcription factor whose expression is limited to the B-cell lineage, developing nervous system and adult testis (Adams et al. 1992). It is important for the transcription of several genes involved in B-cell function, e.g., Ig-α (Fitzsimmons et al. 1996) and CD19 (Kozmik et al. 1992) and also for transcription of GL ε RNA (Rothman et al. 1991; Liao et al. 1994; Thienes et al. 1997; Qiu and Stavnezer 1998). BSAP also binds to sites in the IgH 3′ enhancer and inhibits enhancer activity in transfection assays (Singh and Birshtein 1993; Neurath et al. 1994; Neurath et al. 1995; Singh and Birshtein 1996). BSAP binds to one or two sites 5′ to or within every mouse S region, although the roles of most of these sites are unknown (Waters et al. 1989; Liao et al. 1992; Xu et al. 1992). Targeted deletion of the gene for BSAP ablates the B lineage (Urbanek et al. 1994), and BSAP is expressed in pro-B, pre-B and mature B cells but not in terminally differentiated plasma cells (Waters et al. 1989; Barberis et al. 1990).

In order to attempt to determine whether BSAP regulates CSR, splenic B cells were treated with an antisense oligo for BSAP (Wakatsuki et al. 1994). Although CSR was greatly inhibited, so was cell proliferation and, thus, the reason for the lack of switching could not be determined. More recently, it was found that over-expression of BSAP under control of a tetracycline-regulated operator in I.29μ B cells stimulates transcription from the GL ε promoter and switching to IgE, but also inhibits GL α transcripts and switching to IgA (Qiu and Stavnezer 1998). The proliferation of I.29μ cells is not affected by over-expression of BSAP. Thus, BSAP can regulate the specificity of isotype switching by differential regulation of GL α and ε transcripts. The effect of over-expression of BSAP on GL γ transcripts and switching to IgG has not been reported. BSAP has previously been shown to have the ability to activate or inhibit transcription, apparently depending on which proteins it interacts with at the particular DNA-binding site, although how this is achieved is unknown (Dorfler and Busslinger 1996; Wallin et al. 1998). The binding site for BSAP in the GL ε promoter is a high-affinity site and the putative site in the GL α promoter is a low affinity site, so levels of BSAP which stimulate GL ε transcription may not be sufficient to inhibit GL α transcripts.

### 3.8 Prostaglandin $E_2$

Prostaglandin $E_2$ ($PGE_2$) has been shown to increase GL ε transcripts and IgE and IgG1 synthesis in mouse-splenic B cells activated with LPS + IL-4 (Roper et al. 1990, 1995). The mechanism of $PGE_2$ action is likely to be at least partly due to its ability to induce cAMP, as cAMP increases the production of IgE and IgG1 in LPS + IL-4-activated splenic B cells (Roper et al. 1990). Furthermore, cholera toxin, another inducer of cAMP, also synergizes with LPS and IL-4 to induce GL γ1 transcripts and switching to IgG1 (Lycke et al. 1990). It has recently been found that engagement of CD40 present on human-lung fibroblasts induces $PGE_2$ synthesis by inducing cyclooxygenase-2, a rate-limiting enzyme for $PGE_2$ production (Zhang et al. 1998). Thus, this pathway may be important for induction of IgE synthesis during inflammation of the lungs during allergy and asthma. Induction of $PGE_2$ in IL-4 knock-out mice infected with the mouse retrovirus that causes the murine acquired immunodeficiency syndrome has been invoked to explain the IgE produced in these mice (Morawetz et al. 1996).

### 3.9 CD30

CD30 is a receptor whose engagement has a negative effect on class switching to all isotypes in human B cells (Cerutti et al. 1998). CD30 is a member of the TNF receptor superfamily, which is present on a small population of B cells surrounding the germinal centers of secondary lymphoid organs. It can also be induced on $sIgD^+$ B cells isolated from human peripheral blood lymphocytes by treatment with CD40L (Cerutti et al. 1998). The ligand for CD30, CD30L, is expressed mainly on activated $CD8^+$ T cells (Smith et al. 1993). Treatment of B cells with antibody to CD30 completely inhibits induction of all GL γ, ε and α transcripts by CD40L (Jumper et al. 1995; Cerutti et al. 1998). The inhibition of GL transcripts is at least partially explained by the fact that CD30 signaling inhibits activation of NF-κB by CD40 engagement (Lee et al. 1997). As described above for the mouse GL γ1 promoter, the human GL γ3 promoter has tandem binding sites for NF-κB that drive transcription of a reporter gene in response to CD40L (Cerutti et al. 1998). Antibody to CD30 inhibits both the binding of NF-κB to these sites and transcription driven by the promoter, although the inhibition is not as complete as inhibition of the GL γ3 transcripts themselves.

## 4 Additional Levels of Regulation of CSR

### 4.1 Interleukin-4

IL-4 increases switch recombination at other levels besides regulation of GL transcripts, but the mechanism(s) for this is not yet understood. The I.29μ B-cell line is induced by LPS to undergo CSR to IgA, and addition of TGF-β1 further increases switching to IgA by five- to tenfold by the induction of GL α transcripts (SHOCKETT and STAVNEZER 1991). Antibody to IL-4 inhibits switching to IgA in these cells by 80%, although it has no effect on expression of GL α transcripts or on DNA synthesis. The CH12.LX B-cell line and its subclone CH12F3 can be induced to switch to IgA with IL-4 alone without altering GL α transcript levels (WHITMORE et al. 1991; NAKAMURA et al. 1996). IL-4 is also important for optimal switching to IgA in the system described by MCINTYRE et al. (1995), in which splenic B cells are treated with LPS + TGF-β1 + anti-δ dex + IL-5 + IL-4. Although IL-4 is known to contribute to B-cell proliferation in cultured splenic B cells, it does not increase proliferation or GL α transcripts in the McIntyre system. Furthermore, IL-4-deficient mice have some deficiencies in IgA responses to oral immunization (VAJDY et al. 1995). These data suggest that IL-4 has another role in the class switch to IgA besides induction of GL transcripts or proliferation. It is unknown if this effect is unique to switching to IgA.

## 4.2 Interleukin-5

IL-5 is required in order to obtain CSR to IgG1 in mouse-splenic B cells treated with anti-δ dex + IL-4 (MANDLER et al. 1993a). IL-5 is also required for CSR to IgG1 and to IgE by splenic B cells treated with anti-IgM conjugated to Sepharose beads + IL-4 (PURKERSON and ISAKSON 1992). In both cases, IL-5 was shown to not alter the levels of GL transcripts. These data allowed MANDLER et al. (1993a) to propose that a minimum of three components are required for induction of CSR. They proposed that CSR requires accessibility of the switch regions, cell proliferation (induced by anti-δ dex or anti-IgM coupled to Sepharose beads) and some additional activity, perhaps switch recombinase, induced in this model by IL-5. In support of this model, incubation of splenic B cells with anti-δ dex in the presence of IL-5, but not in its absence, has been found to induce one or more specific double-strand breaks in the Sγ3 region (WUERFFEL et al. 1997).

By contrast, IL-5 is not required to induce switching in splenic B cells incubated with LPS, with a form of CD40L which cross-links CD40 or with antibody to CD40. MANDLER et al. (1993a) hypothesized that these activators induce the putative switch recombinase in addition to inducing proliferation. LPS alone induces the same double-strand break(s) in Sγ3 as does anti-δ dex + IL-5 (WUERFFEL et al. 1997). The distinction between the effects of B-cell activators on proliferation from their effects on recombination can also be observed in experiments in which B-cell lines are induced to undergo CSR. For example, in the presence of TGF-β1, I.29μ cells are induced to switch by LPS, although LPS does not induce DNA synthesis or proliferation in I.29μ cells (SHOCKETT and STAVNEZER 1991).

IL-5 is also required for optimal IgA switching in splenic B cells induced with LPS (or CD40L) + anti-δ dex + IL-4 + TGF-β1 (MCINTYRE et al. 1995). It is unknown why so many signals are required for optimal switching to IgA.

## 4.3 Interleukin-10

IL-10 regulates switching to IgG3 and IgA in mouse B cells at another level besides regulation of GL transcripts, as discussed above in Sect. 3.4.4 (Shparago et al. 1996).

## 4.4 CD40L and LPS

The activators CD40L and LPS induce NF-κB, which, as described above, directs CSR by regulating GL transcripts. They also induce cell proliferation and induce CSR at another unknown level(s). This level of induction of proliferation and CSR by CD40L does not require NF-κB p50, although induction by LPS does (Sha et al. 1995; Snapper et al. 1996b).

## 4.5 Cell Cycle and Cell Division

CSR occurs only in mouse-splenic B cells that have been activated to proliferate and treatment of cells with inhibitors of DNA synthesis prevents CSR (Severinson-Gronowicz et al. 1979; Kenter and Watson 1987). Furthermore, a specific number of cell divisions appears to be required, since IgG1 is expressed only on the surface of cells which have undergone at least three rounds of cell division, and IgE is only expressed on cells which have undergone at least six rounds (Hodgkin et al. 1996; Hasbold et al. 1998). Interestingly, the same number of cell divisions is required for IgG1 switching, no matter whether cells are stimulated with optimal or sub-optimal amounts of IL-4. After the third division cycle, cells in the population begin to express sIgG1, and most cells actually undergo additional rounds of proliferation prior to expressing IgG1. The requirement for additional replications for IgE expression is in accord with previous findings that switching to IgE in splenic B cells mostly occurs via first switching to IgG1 (Siebenkotten et al. 1992; Mandler et al. 1993b).

It is entirely unclear why three rounds of division must occur before IgG1 is expressed, but it suggests that a series of inductions need to occur and that one or more of these steps require replication of DNA. Possible steps might include induction of chromatin accessibility by binding of specific transcription factors, followed by histone acetylation, followed by DNA demethylation, followed by transcription of GL transcripts and formation of a recombinase complex, recombination, repair of DNA breaks and synthesis of the new heavy-chain mRNA and antibody. Since recombination itself has not been examined in these experiments, it is possible that CSR may occur after two rounds of cell division. The finding that a specific number of cell divisions must occur prior to CSR is not unique to processes involving DNA recombination, since it has recently been shown that specific numbers of cell divisions are required for production of different cytokines by $T_H$ cells (Gett and Hodgkin 1998).

Although the results mentioned above indicate that CSR only occurs in proliferating cells, these data do not indicate at which point in the cell cycle it occurs. By using elutriation to separate splenic B cells at different stages of the cell cycle, it has been shown that expression of GL $\gamma 1$ transcripts is regulated during the cell cycle, being at the highest levels in the G1 and S phases and at reduced levels in the G2 and M phases and absent in G0 (LUNDGREN et al. 1995). Using EMSAs, two protein complexes that bind an Ets site in the promoter for GL $\gamma 1$ transcripts and that are regulated similarly to the GL $\gamma 1$ RNA levels were identified. Evidence suggested that during the G2 and M phases these complexes leave the DNA and then reassemble during the subsequent G1 phase. Since it appears likely that GL transcripts must be present at the time of CSR, these data suggest that CSR occurs during G1 and/or early S phase. After recombination, the DNA breaks would need to be repaired in order to successfully complete S phase and duplicate the chromosomes.

Also consistent with the hypothesis that CSR occurs in G1 or early S phase is the finding that sister-chromatid exchange does not occur during CSR (WABL et al. 1985). If CSR occurred in G2 phase when the two sister chromatids are aligned, it is likely that sister-chromatid exchange would be observed. In conclusion, the available data suggest that CSR occurs in G1 and is complete prior to S phase, although there are no direct data supporting this hypothesis.

# 5 Model for Switch Recombination

## 5.1 Double-Strand Breaks, Ku, Illegimate Priming and End-Joining

Antibody class switching occurs by recombination between S regions. S regions consist of G-rich, simple tandem repetitive sequences 1–10kb in length. The repeat units vary from 20 to 80 nucleotides in length, depending on the isotype (reviewed by GRITZMACHER 1989). Although the switch regions for each isotype differ from each other, they contain small elements in common, e.g., GAGCT, GGGGT and GGGCT. Switch recombination can occur at many different sites within S regions, perhaps at every nucleotide, and there is no precise sequence specificity for the site of recombination within the S regions. Several motifs have been defined within S regions that appear to be preferentially found near switch junctions at statistically significant levels. However, no motif occurs consistently, even within the same heavy-chain class or subclass, and no motif occurs at the same position relative to the break (KENTER and BIRSTEIN 1981; MARCU et al. 1982; CHOU and MORRISON 1993; DUNNICK et al. 1993; KENTER et al. 1993).

Switch recombination results in excision of a DNA circle containing the chromosomal segment residing between the two recombined switch regions (Fig. 1; IWASATO et al. 1990; MATSUOKA et al. 1990; VON-SCHWEDLER et al. 1990). This finding indicates that the initiating event in switch recombination must result in two double-stranded breaks occurring in a concerted fashion, although the actual breakage event could occur by sequential single-strand breaks that would not need to be directly across from each other. Interestingly, a double-strand break has been shown to be rapidly induced in one of the tandem repeats in the Sγ3 region in splenic B cells treated with LPS or with anti-δ dex + IL-5, but not by anti-δ dex alone, i.e., it is found in conditions that induce CSR to IgG3 (WUERFFEL et al. 1997). Only one clearly specific inducible break point and one likely additional break have been detected thus far, but since only three tandem Sγ3 repeats have been examined, many more breaks may be induced.

The double-strand break is present at the earliest time point tested, 4h after activation. This break is maintained until the last time point tested, 44h. It is unknown whether GL transcription is required to obtain the break, since under all conditions examined GL γ3 transcripts are present. The data indicate, however, that GL transcripts are not sufficient to induce the double-strand break, since GL γ3 transcripts are present in cells treated with anti-δ dex alone, conditions that do not induce the break (WUERFFEL et al. 1997). The regulation of the double-strand endonuclease should be a key step in regulating CSR.

### 5.1.1 Role of Ku

Consistent with the involvement of double-strand breaks in initiation of CSR, DNA-dependent protein kinase (DNA-PK), Ku70 and Ku80 have been shown to be required for CSR by targeted mutation of each of these genes in mice (ROLINK et al. 1996; CASELLAS et al. 1998; MANIS et al. 1998a). These proteins are known to form a complex and are involved in double-strand-break repair and end-joining reactions (RATHMELL and CHU 1994a,b; LIANG et al. 1996). All three proteins are required for V(D)J recombination (TACCIOLI et al. 1993; BLUNT et al. 1995).

A complex of Ku70 and Ku80 that binds DNA is induced in nuclear extracts from splenic B cells treated with CD40L and IL-4, as assayed by EMSAs (ZELAZOWSKI et al. 1997b). Neither CD40L nor IL-4 alone induce the complex. Binding of the Ku complex is also induced by anti-δ dex, but IL-4 does not further induce it. IL-5 has little or no effect on the induction by anti-δ dex. It is possible that the induction of Ku is part of the role that CD40L and IL-4 and anti-δ dex play in induction of CSR. However, since both anti-δ dex and the combination of CD40L and IL-4 are highly stimulatory for B-cell proliferation (BRUNSWICK et al. 1988; ROUSSET et al. 1991), it is also possible that induction of DNA binding by Ku has a role in maintaining chromosomal integrity during cell proliferation that is not specific for CSR. Furthermore, it is possible that the role of Ku may be to prevent cell death due to an inability to repair chromosomes cut by the putative endonuclease(s) that initiates CSR.

Point mutations, insertions and deletions occur near the switch-recombination junctions, indicating that an error-prone DNA synthesis mechanism is part of the process of CSR (DUNNICK et al. 1989, 1993). Because of this finding, an illegitimate priming model for switch recombination which involves priming DNA synthesis by one switch region, e.g., Sμ, on a downstream switch region, e.g., Sε, has been proposed (Fig. 4; DUNNICK et al. 1993; DUNNICK and STAVNEZER 1990). Simultaneously, the downstream switch region primes DNA synthesis on the upstream switch region (Fig. 4). According to this model, after recombination and segregation of the products into two different daughter cells, one of the two daughter cells has a switch junction containing a newly synthesized, mutated, downstream switch region and unmutated, parental, upstream switch sequence (Fig. 4). The switch junction in the other daughter cell has a newly synthesized (mutated) upstream switch sequence and parental downstream switch region. Although many examples of switch recombination junctions with mutations have been obtained, in most cases the sequences of the parental switch regions on both sides of the junctions are not known, thus preventing analysis for mutations on both sides of the junction. However, in the few cases in which this has been possible, nearly all the mutations are found on only one side of the switch junction, which is consistent with the illegitimate priming model (DUNNICK et al. 1993; DU et al. 1997).

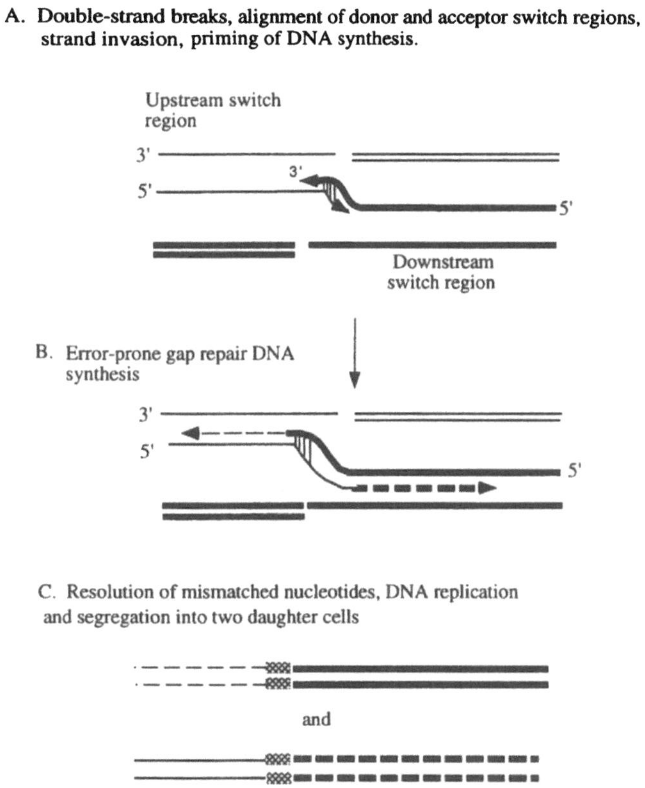


**Fig. 4.** Illegitimate-priming model for class-switch recombination (from Dunnick et al. 1990, 1993; Stavnezer 1996b). See text for further description. In this diagram, the strands that do not participate in the chromosomal recombination event shown recombine to form the excised circle that is deleted from the chromosome

The illegitimate priming model might seem to predict the finding of sequence identities between the upstream and downstream switch regions at the switch junctions. Although such identities are commonly found in switch recombination, there is no preference for junctions with long identities (DUNNICK et al. 1993). Thus, switch recombination differs from homologous recombination. In agreement with this, mice with a targeted mutation in the Rad54 gene have a reduced ability to undergo homologous recombination but undergo normal levels of class switching (ESSERS et al. 1997). The finding of short bits of identity at recombination junctions is typical of illegitimate recombination, or end-joining, in mammals (ROTH and WILSON 1986). It is also typical of V(D)J recombination (LEWIS 1994). An interesting observation that is consistent with the hypothesis that error-prone DNA synthesis creates the switch junction and also that CSR occurs by an end-joining (illegitimate) mechanism has been reported by LEHMAN et al. (1994). This group showed that recombination by end-joining in germinal vesicle extracts from *Xenopus* oocytes requires deoxynucleotide triphosphates and, thus, appears to involve DNA synthesis. By contrast, this same group has found that homologous recombination in these extracts does not require deoxynucleotide triphosphates (LEHMAN and CARROLL 1991).

### 5.1.2 Rad51

CSR has been postulated to involve Rad51, an enzyme involved in homologous recombination and double-strand-break DNA repair, and whose deletion results in death at the four-cell-stage in mice (LIM and HASTY 1996). This suggestion was made because Rad51 mRNA and protein are expressed between 2 days and 5 days after immunization in localized foci in the spleen that also contain GL γ1 transcripts and IgM and IgG (LI et al. 1996; LI and MAIZELS 1997; PEAKMAN and MAIZELS 1998). Since CSR is accompanied by a high rate of cell proliferation, it is entirely possible that Rad51 is involved in other functions required during chromosomal replication occurring in B cells activated during an immune response.

CSR does not require Rag-1 and Rag-2, the lymphoid-specific enzymes which recognize V(D)J-recombination signal sequences and initiate the recombination by nicking the DNA and creating the hairpin. This was demonstrated by creating mice lacking both Rag-1 and Rag-2 but having a rearranged expressed μ gene inserted into the heavy-chain locus and an expressed transgenic λ gene. B cells from these mice responded similarly to control mice to LPS or LPS + IL-4 to undergo CSR (LANSFORD et al. 1998).

## 5.2 Proteins that Bind the S-Region Consensus Repeats

### 5.2.1 NF-κB p50 and E47

Kenter and colleagues analyzed the consensus repeats of Sγ3, Sγ1 and Sγ2b for localization of sites of recombination with Sμ and found that switch recombination occurs more frequently in the first 30 bp of the 49-bp repeats of these Sγ regions than expected by chance (Wuerffel et al. 1992; Kenter et al. 1993; Ma et al. 1997). Sites of recombination for Sγ2a have not been determined. They have also found that within these 30-bp regions there are binding sites for NF-κB/p50 and a complex (switch nuclear A protein) SNAP, which is inducible by LPS + dextran sulfate and contains an E47-like protein (Wuerffel et al. 1992; Ma et al. 1997). E47 is a helix-loop-helix transcription factor that binds the E2 box of the μ-intron enhancer and also binds the promoters of several genes (Murre and Baltimore 1992; Ma et al. 1997). Kenter et al. hypothesize that both the NF-κB/p50 and SNAP complexes are important for specifying sites of switch recombination. It is unknown if Sε and Sα regions also bind NF-κB/p50 or complexes containing E47.

It is clear, however, that CSR to IgG subclasses can occur in p50-deficient mice (Sha et al. 1995; Snapper et al. 1996a,b; Zelazowski et al. 1997a). As described above, some of the isotype-specific effects of deletion of the p50 gene can be explained by the effect of NF-κB on transcription of their respective GL RNAs, in addition to the finding that p50-deficient B cells cannot proliferate in response to LPS (Sha et al. 1995; Lin and Stavnezer 1996; Lin et al. 1998). However, as also described above, NF-κB appears to have additional roles in CSR to IgA and IgE that cannot be explained by effects on GL transcripts and cell proliferation. It will be interesting to determine whether the sites of Sμ-Sγ switch recombination in p50-deficient mice are altered. A targeted mutation resulting in ablation of the E47-coding region of the E2A gene greatly inhibits development of B cells, but the effect on CSR in these mice has not been reported (Bain et al. 1997).

### 5.2.2 LR1

LR1 is a transcription factor which regulates transcription of the *c-myc* gene (Brys and Maizels 1994) and which also binds a site in Sγ1, a site in the Ig μ intron enhancer and possibly a site in Sγ3 (Williams and Maizels 1991; Williams et al. 1993). Additional binding sites in switch regions have been postulated. LR1 is a B-cell-specific protein with the same B-cell stage-specific distribution as BSAP/Pax5. LR1 has been demonstrated in EMSA to form a complex with an abundant nucleolar protein called nucleolin or C23 (Hanakahi et al. 1997). Nucleolin is involved in the processing of ribosomal RNA. In independent experiments, nucleolin has also been shown to be part of a complex that contains a protein (nucleophosmin/B23) that is capable of forming joint molecules between a 49-nucleotide single-stranded Sγ oligonucleotide and a plasmid containing a 1.3-kb Sμ sequence (Borggrefe et al. 1998). Formation of the joint molecules does not occur if the plasmid is devoid of S-region sequences. The complex also contains poly(adenosine diphosphate-ribose) polymerase, an enzyme involved in DNA repair, whose inhibition in I.29μ cells stimulates CSR to IgA (Shockett and Stavnezer 1993), but whose deletion from the mouse genome by gene targeting has no effect on switching to IgA in cultured splenic B cells (Wang et al. 1997). The complex also contains a novel protein of unknown function, SWAP-70, which is induced in B cells treated with LPS or CD40L + IL-4 (BORGGREFE et al. 1998).

## 6 The IgH 3′ Enhancer Regulates GL Transcripts and Switching

The IgH 3′ enhancer, located 16kb 3′ to the Cα gene, appears to be involved in regulating class switching. Substitution of a portion of the IgH 3′ enhancer with a neo$^r$ gene driven by a phosphoglycerate-kinase (PGK) promoter results in pronounced defects in switching to most isotypes, although such mice have normal numbers of otherwise-normal B cells (COGNE et al. 1994). Switching to IgG3, IgG2b, IgG2a and IgE is greatly reduced, but not to IgG1 and IgA in B cells activated in culture. The levels of GL transcripts for the isotypes that are inhibited are reduced, suggesting that the effect is due to inhibition of expression of GL transcripts.

How the IgH 3′ enhancer mutation inhibits GL transcripts is unknown, although MADISEN and GROUDINE (1994) have proposed that it serves as a locus-control region (LCR), since it has copy-number-dependent enhancer activity. The finding that the levels of several GL transcripts are reduced by the replacement of elements of the IgH 3′ enhancer with the PGK-neo$^r$ gene also supports the hypothesis that it is a LCR. Replacement of elements within the LCR for the β-globin gene cluster with a PGK-neo$^r$ gene has been shown to inhibit expression of several genes within the β-globin gene cluster (FIERING et al. 1995; HUG et al. 1996). If the same elements of the β-globin LCR are simply deleted from the chromosome, the effects on expression of the genes are minimal. This is also true for the IgH 3′ enhancer. If a portion of it is deleted rather than substituted with the PGK-neo$^r$ gene, there are no effects on CSR (MANIS et al. 1998b). A similar finding has been obtained for the 3′ Ig-κ enhancer. Replacement of the 3′ Ig-κ enhancer with the PGK-neo$^r$ gene inhibits development of $\kappa^+$ B cells to a greater extent than does deletion of the 3′ κ enhancer (GORMAN et al. 1996). The PGK promoter is hypothesized to have a special ability to disrupt the function of LCRs.

A related result has been obtained in B cells from mice in which the Iα exon and proximal promoter region have been replaced with a PGK-mini-HPRT gene (QIU et al. 1999). As described above, LPS-activated B cells from these mice switch to IgA without treatment with TGF-β1. Interestingly, when switching to IgG in these B cells was examined, it was found that switching to all four IgG subclasses is reduced by from 45% to over 90% in B cells activated by either LPS- or anti-δ dextran + IL-5. The authors suggest that replacement of the Iα exon with the PGK-HPRT gene inhibits expression of GL γ transcripts, perhaps by sequestering the LCR and thereby inhibiting interaction of the promoters for these transcripts with the LCR. The levels of GL γ transcripts were not examined, however. Experiments of Seidl and Alt (1999) extend these findings, as they report that replacement of the Iγ2b exon with a PGK-neo$^r$ gene in the correct orientation results in inhibition of switching to IgG3 and IgG1, but not to any of the isotypes whose $C_H$ genes are located 3′ to Cγ2b. These results suggest that the PGK promoter may sequesten the enhancer, preventing access to the LCR by GL RNA promoters located upstream of the PGK-neo$^r$ gene.

The inhibition of switching to other isotypes by substitution of the PGK-HPRT or PGK-neo$^r$ genes for the Iα or Iγ2b exons, respectively, may be specific to the PGK promoter. Three examples have been reported in which a neo$^r$ gene driven by the polyoma enhancer/herpes virus thymidine kinase promoter replaced an I exon (Zhang et al. 1993; Bottaro et al. 1994; Jung et al. 1994). B cells from these mice showed no alteration in ability to switch to any of several isotypes, except for the one corresponding to the targeted I exon. However, since in these three cases the neo$^r$ gene was inserted in antisense polarity, it is still possible that the lack of inhibition could be due to the polarity of the neo$^r$ gene.

# 7 Concluding Remarks

I will conclude by mentioning some of the interesting questions that need to be addressed and also a few speculations about the mechanism of CSR. There is still much to be learned about switch recombination, most notably the function of GL transcripts and the mechanism of the recombination event itself. Also, why does the GL transcript need to be spliced in order to obtain switch recombination? It is clear that isotype specificity is regulated by GL transcripts, but it is unknown whether there are class-specific switch recombinase activities that also regulate isotype specificity.

Although the finding of a double-strand break induced in the Sγ3 region during induction of switch recombination is very appealing as an initial step in the recombination process, it is unknown whether this is truly an initiating event. More importantly, it is quite possible there are numerous DNA breaks and only a subset of them go on to generate switch recombination. Since anti-δ dextran induces both DNA replication and transcription of GL γ3 RNA but does not induce double-strand breaks, it appears unlikely that the role of GL transcripts is to direct the putative endonuclease to particular switch regions. It appears more likely that the transcripts would act later during the recombination process, perhaps aiding the strand alignment and strand invasion that probably occur during the search for the bits of homology at which the joining occurs (Stavnezer 1996a). Müller et al. (1998) have proposed that the GL transcripts serve as templates for reverse transcription primed by the Sμ DNA sequence and that this creates the switch junction. Are Ku and DNA-PK required for the recombination event itself? Or are they required simply to maintain cell viability by aiding in the massive DNA repair that would be required if induction of class switching results in induction of a large number of DNA breaks?

Another series of interesting questions has arisen from the findings that IL-4 and IL-5, CD40L, LPS and NF-κB p50 all contribute to CSR at levels other than simply induction of GL transcripts and DNA synthesis. What are their roles? Also, why does it require three rounds of cell division to detect expression of IgG1 on B cells after inducers of switch recombination have been added?

*Acknowledgements.* I thank Drs. Kathy Seidl and Fred Alt for sharing their unpublished data and Dr. Amy Kenter for many helpful discussions. This publication was made possible by grants AI-23287 and AI-42108 from the National Institutes of Health. Its contents are solely the responsibility of the author and do not necessarily represent the official views of the National Institutes of Health.

## References

Adams B, Dorfler P, Aguzzi A, Kozmik ZY, Urbanek P, Maurer-Fogy I, Busslinger M (1992) Pax-5 encodes the transcription factor BSAP and is expressed in B lymphocytes, the developing CNS, and adult testis. Genes Dev 6:1589–1607

Akimoto T, Numata F, Tamura M, Higashida N, Takashi T, Takeda K, Akira S (1998). Abrogation of bronchial eosinophilic inflammation and airway hyperreactivity in STAT6-deficient mice. J Exp Med 187:1537–1542

Albrecht B, Peiritsch S, Woisetschlager M (1994) A bifunctional control element in the human IgE germline promoter involved in repression and IL-4 activation. Int Immunol 6:1143–1151

Allen RC, Armitage RJ, Conley ME, Rosenblatt H, Jenkins NA, Copeland NG, Bedell MA, Edelhoff S, Disteche CM, Simoneaux DK, Fanslow WC, Belmont J, Spriggs MK (1993) CD40 ligand gene defects responsible for X-linked hyper-IgM syndrome. Science 259:990–993

Andersson J, Coutinho A, Melchers F (1978) The switch from IgM to IgG secretion in single mitogen-stimulated B-cell clones. J Exp Med 147:1744–1754

Aruffo A, Farrington M, Hollenbaugh D, Li X, Milatovich A, Nonoyama S, Bajorath J, Grosmaire L, Stenkamp R, Neubauer M, Roberts R, Noelle R, Ledbetter J, Franke U, Ochs H (1993) The CD40 ligand, gp39, is defective in activated T cells from patients with X-linked hyper-IgM syndrome. Cell 72:291–300

Bain G, Maandag ECR, Teriele HP, Feeney AJ, Sheehy A, Schlissel M, Shinton SA, Hardy RR, Murre C (1997) Both E12 and E47 allow commitment to the B cell lineage. Immunity 6:145–154

Barberis A, Widenhorn K, Vitelli L, Busslinger M (1990) A novel B-cell lineage-specific transcription factor present at early not late stages of differentiation. Genes Dev 4:849–859

Beg AA, Sha WC, Bronson RT, Ghosh S, Baltimore D (1995) Embryonic lethality and liver degeneration in mice lacking the RelA component of NF-κB. Nature 376:167–170

Berberich I, Shu GL, Clark EA (1994) Cross-linking CD40 on B cells rapidly activates nuclear factor-κB. J Immunol 153:4357–4366

Bergstedt-Lindqvist S, Moon H-B, Persson U, Moller G, Heusser C, Severinson E (1988) Interleukin 4 instructs uncommitted B lymphocytes to switch to IgG1 and IgE. Eur J Immunol 18:1073–1077

Bergstedt-Lindqvist S, Sideras P, Macdonald HR, Severinson E (1984) Regulation of Ig class secretion by soluble products of certain T-cell lines. Immunol Rev 78:25–50

Berton MT, Linehan LA (1995) IL-4 activates a latent DNA-binding factor that binds a shared IFN-γ and IL-4 response element present in the germ-line γ1 Ig promoter. J Immunol 154:4513–4525

Berton MT, Uhr JW, Vitetta ES (1989) Synthesis of germline γ1 immunoglobulin heavy-chain transcripts in resting B cells: induction by interleukin 4 and inhibition by interferon γ. Proc Natl Acad Sci USA 86:2829–2833

Berton MT, Vitetta ES (1990) Interleukin 4 induces changes in the chromatin structure of the γ1 switch region in resting B cells before switch recombination. J Exp Med 172:375–378

Berton MT, Vitetta ES (1992) IL-4-induced expression of germline γ1 transcripts in B cells following cognate interactions with T helper cells. Int Immunol 1992:387–396

Blunt T, Finnie NJ, Taccioli GE, Smith GCM, Demengeot J, Gottlieb TM, Mizuta R, Varghese AJ, Alt FW, Jeggo PA, Jackson SP (1995) Defective DNA-dependent protein kinase activity is linked to

V(D)J recombination and DNA repair defects associated with the murine scid mutation. Cell 80: 813–823
Borggrefe T, Wabl M, Akhmedov AT, Jessberger R (1998) A B-cell specific DNA recombination complex. J Biol Chem 273:17025–17035
Bottaro A, Lansford R, Xu L, Zhang J, Rothman P, Alt F (1994) I region transcription (per se) promotes basal IgE class switch recombination but additional factors regulate the efficiency of the process. EMBO J 13:665–674
Brunswick M, Finkelman FD, Highet PF, Inman JK, Dintzis HM, Mond JJ (1988) Picogram quantities of anti-Ig antibodies coupled to dextran induce B cell proliferation. J Immunol 140:3364–3372
Brys A, Maizels N (1994) LR1 regulates c-myc transcription in B-cell lymphomas. Proc Nat Acad Sci USA 91:4915–4919
Callard RE, Matthews DJ, Hibbert LM (1997) IL-4 and IL-13: same response, different receptors. Biochem Soc Trans 25:451–455
Carroll MC (1998) The role of complement and complement receptors in induction and regulation of immunity. Annu Rev Immunol 16:545–568
Casellas R, Nussenzweig A, Wuerffel R, Pelanda R, Reichlin A, Suh H, Qin X-F, Besmer E, Kenter A, Rajewsky K, Nussenzweig MC (1998) Ku80 is required for immunoglobulin isotype switching. EMBO J 17:2404–2411
Castigli E, Alt FW, Davidson L, Bottaro A, Mizoguchi E, Bhan AK, Geha RS (1994) CD40-deficient mice generated by recombination-activating gene-2-deficient blastocyst complementation. Proc Nat Acad Sci USA 91:12135–12139
Cerutti A, Schaffer A, Shah S, Zan H, Liou H-C, Goodwin RG, Casalli P (1998) CD30 is a CD40-inducible molecule that negatively regulates CD40-mediated immunoglobulin class switching in non-antigen-selected human B cells. Immunity 9:247–256
Choi MS, Brines RD, Holman MJ, Klaus GG (1994) Induction of NF-AT in normal B lymphocytes by anti-immunoglobulin or CD40 ligand in conjunction with IL-4. Immunity 1:179–197
Chou CL, Morrison SL (1993) A common sequence motif near nonhomologous recombination breakpoints involving Ig sequences. J Immunol 150:5350–5360
Chu CC, Max EE, Paul WE (1993) DNA rearrangement can account for in vitro switching to IgG1. J Exp Med 178:1381–1390
Chu CC, Paul WE, Max EE (1992) Quantitation of immunoglobulin μ-γ1 heavy chain switch region recombination by a digestion-circularization polymerase chain reaction method. Proc Nat Acad Sci USA 89:6978–6982
Coffman RL, Lebman DA, Rothman PB (1993) Mechanism and regulation of immunoglobulin isotype switching. Adv Immunol 54:229–270
Coffman RL, Lebman DA, Shrader B (1989) Transforming growth factor-β specifically enhances IgA production by lipopolysaccharide-stimulated murine B lymphocytes. J Exp Med 170: 1039–1044
Coffman RL, Ohara J, Bond MW, Cary J, Zlotnik A, Paul WE (1986) B cell stimulatory factor-1 enhances the IgE responses of lipopolyasaccharide-activated B cells. J Immunol 136:4538–4545
Cogne M, Lansford R, Bottaro A, Zhang J, Gorman J, Alt FW (1994) A class switch control region at the 3′ end of the immunoglobulin heavy chain locus. Cell 77:737–747
Collier DA, Griffin JA, Wells RD (1988) Non-B right-handed DNA conformations of homopurine-homopyrimidine sequences in the murine immunoglobulin Cα switch region. J Biol Chem 263: 7397–7405
Collins JT, Dunnick WA (1993) Germline transcripts of the murine immunoglobulin γ2a gene: structure and induction by IFN-γ. Int Immunol 5:885–891
Cooke MP, Heath AW, Shokat KM, Zheng Y, Finkelman FD, Linsley PS, Howard M, Goodnow CC (1994) Immunoglobulin signal transduction guides the specificity of B cell-T cell interactions and is blocked in tolerant self-reactive B cells. J Exp Med 179:425–434
Cooper CL, Berrier AL, Roman C, Calame KL (1994) Limited expression of C/EBP family proteins during B lymphocyte development: negative regulator Ig/EBP predominates early and activator NF-IL6 is induced later. J Immunol 153:5049–5058
Cunningham K, Ackerly H, Alt F, Dunnick W (1998) Potential regulatory elements for germline transcription in or near murine Sγ1. Int Immunol 10:527–536
Darnell JE (1997) STATs and gene regulation. Science 277:1630–1635
Defrance T, Vanbervliet B, Brière F, Durand I, Rousset F, Banchereau J (1992) Interleukin 10 and transforming growth factor β cooperate to induce anti-CD40-activated naive human B cells to secrete immunoglobulin A. J Exp Med 175:671–682

Delphin SA, Stavnezer J (1995) Characterization of an IL-4 responsive region in the immunoglobulin heavy chain ε promoter: Regulation by NF-IL4, a C/EBP family member and NF-κB/p50. J Exp Med 181:181–192
DiSanto JP, Bonnefoy JY, Gauchat JF, Fischer A, de-Saint-Basile G (1993) CD40 ligand mutations in X-linked immunodeficiency with hyper-IgM. Nature 361:541–543
Doi TS, Takahashi T, Taguch O, Azuma T, Obata Y (1997) NF-κB RelA-deficient lymphocytes: normal development of T cells and B cells, impaired production of IgA and IgG1 and reduced proliferative responses. J Exp Med 185:953–961
Dorfler P, Busslinger M (1996) C-terminal activating and inhibitory domains determine the transactivation potential of BSAP(Pax-5), Pax-2 and Pax-8. EMBO J 15:1971
Du J, Shanmugam A, Kenter AL (1997) Analysis of immunoglobulin Sγ3 recombination breakpoints by PCR: implications for the mechanism of isotype switching. Nucl Acids Res 25:3066–3073
Dunnick W, Hertz GZ, Scappino L, Gritzmacher C (1993) DNA sequences at immunoglobulin switch region recombination sites. Nucl Acids Res 21:365–372
Dunnick W, Stavnezer J (1990) Copy choice mechanism of immunoglobulin heavy chain switch recombination. Mol Cell Biol 10:397–400
Dunnick W, Wilson M, Stavnezer J (1989) Mutations, duplication, and deletion of recombined switch regions suggest a role for DNA replication in the immunoglobulin heavy-chain switch. Mol Cell Biol 9:1850–1856
Esser C, Radbruch A (1989) Rapid induction of transcription of unrearranged Sγ1 switch regions in activated murine B cells by interleukin 4. EMBO J 8:483–488
Essers J, Hendriks RW, Swagemakers SMA, Troelstra C, de Wit J, Bootsma D, Hoeijmakers JHJ, Kanaar R (1997) Disruption of mouse RAD54 reduces ionizing radiation resistance and homologous recombination. Cell 89:195–204
Fanslow WC, Anderson DM, Grabstein KH, Clark EA, Cosman D, Armitage RJ (1992) Soluble forms of CD40 inhibit biologic responses of human B cells. J Immunol 149:655–660
Fearon DT, Locksley RM (1996) The instructive role of innate immunity in the acquired immune response. Science 272:50–53
Fenghao X, Saxon A, Nguyen A, Ke Z, Diaz-Sanchez D, Nel A (1995) Interleukin 4 activates a signal transducer and activator of transcription (Stat) protein which interacts with an interferon-γ activation site-like sequence upstream of the Iε exon in a human B cell line: evidence for the involvement of Janus kinase 3 and interleukin-4 Stat. J Clin Invest 96:907–914
Ferlin WG, Severinson E, Strom L, Heath AW, Coffman RL, Ferrick DA, Howard MC (1996) CD40 signaling induces interluekin-4-independent IgE switching in vivo. Europ J Immunol 26: 2911–2915
Fiering S, Epner E, Robinson K, Zhuang Y, Telling A, Hu M, Martin DIK, Enver T, Ley TJ, Groudine M (1995) Targeted deletion of 5′ HS2 of the murine β-globin LCR reveals that it is not essential for proper regulation of the β-globin locus. Genes Dev 9:2203–2213
Finkelman FD, Katona IMJF, Urban J, Snapper CM, Ohara J, Paul WE (1986) Suppression of in vivo polyclonal IgE respones by monoclonal antibody to the lymphokine B-cell stimulatory factor 1. Proc Nat Acad Sci U S A 83:9675–9679
Finkelman FD, Katona IM, Mosmann TR, Coffman RL (1988) IFN-γ regulates the isotypes of Ig secreted during in vivo humoral immune responses. J Immunol 140:1022–1030
Fitzsimmons D, Hodsdon W, Wheat W, Maira S-M, Wasylyk B, Hagman J (1996) Pax-5 (BSAP) recruits Ets proto-oncogene family proteins to form functional ternary complexes on a B-cell-specific promoter. Genes Dev 10:2198
Foy TM, Aruffo A, Bajorath J, Buhlmann JE, Noelle RJ (1996) Immune regulation by CD40 and its ligand gp39. Ann Rev Immunol 14:591–618
Francis DA, Karras JG, Ke XY, Sen R, Rothstein TL (1995) Induction of the transcription factors NF-κB, AP-1 and NF-AT during B cell stimulation through the CD40 receptor. Int Immunol 7:151–161
Fujieda S, Saxon A, Zhang K (1996) Direct evidence that gamma 1 and gamma 3 switching in human B cells is IL-10 dependent. Mol Immunol 33:1335–1343
Fujieda S, Zhang K, Saxon A (1995) IL-4 plus CD40 mAb induces human B cells gamma subclass specific istoype switch: Switching to γ1, γ3, γ4 but not γ2. J Immunol 155:2318–2328
Fuleihan R, Ramesh N, Loh R, Jabara H, Rosen RS, Chatila T, Fu SM, Stamenkovic I, Geha RS (1993) Defective expression of the CD40 ligand in X chromosome linked immunoglobulin deficiency with normal or elevated IgM. Proc Nat Acad Sci USA 90:2170–2173
Gaff C, Grumont RJ, Gerondakis S (1992) Transcriptional regulation of the germline immunoglobulin Cα and Cε genes: implications for commitment to an isotype switch. Int Immunol 4:1145–1151

Gascan H, Gauchat J-F, Aversa G, van-Vlasselaer P, de Vries JE (1991) Anti-CD40 monoclonal antibodies or CD4+ T cell clones and IL-4 induce IgG4 and IgE switching in purified human B cells via different signaling pathways. J Immunol 147:8–13

Gauchat J-F, Aversa G, Gascan H, de Vries JE (1992) Modulation of IL-4 induced germline ε RNA synthesis in human B cells by tumor necrosis factor-α, anti-CD40 monoclonal antibodies or transforming growth factor-β correlates with levels of IgE production. Int Immunol 4:397–406

Gauchat J-F, Lebmen DA, Coffman RL, Gascan H, deVries JE (1990) Structure and expression of germline ε transcripts in human B cells induced by interleukin 4 to switch to IgE production. J Exp Med 172:463–473

Gearhart PJ, Sigal NH, Klinman NR (1975) Production of antibodies of identical idiotypes but diverse immunogloblin classes by cells derived from a single stimulated B cell. Proc Nat Acad Sci USA 72:1707–1711

Gerondakis S, Gaff C, Goodman D J, Grumont RJ (1991) Structure and expression of mouse germline immunoglobulin γ3 heavy chain transcripts induced by the mitogen lipopolysaccharide. Immunogen 34:392–400

Ghosh S, May MJ, Kopp EB (1998) NF-kappa B and Rel proteins: evolutionarily conserved mediators of immune responses. Annu Rev Immunol 16:225–60

Gilliam AC, Shen A, Richards JE, Blattner FR, Mushinski JF, Tucker PW (1984) Illegitimate recombination generates a class switch from Cμ to Cδ in an IgD-secreting plasmacytoma. Proc Nat Acad Sci U S A 81:4164–4168

Goodman DJ, Gaff C, Gerondakis S (1993) The IL-4 induced increase in the frequency of resting murine splenic B cells expressing germline Ig heavy chain γ1 transcripts correlates with subsequent switching to IgG1. Int Immunol 5:199–208

Gorman JR, Stoep NVD, Monroe R, Cogne M, Davidson L, Alt FW (1996) The Igκ enhancer influences the ratio of Igκ versus Igλ B lymphocytes. Immunity 5:241–252

Goroff DK, Holmes JM, Bazin H, Nisol F, Finkelman FD (1991) Polyclonal activation of the murine immune system by an antibody to IgD. XI. Contribution of membrane IgD cross-linking to the generation of an in vivo polyclonal antibody response. J Immunol 146:18–25

Gritzmacher CA (1989) Molecular aspects of heavy-chain class switching. Crit Rev Immunol 9:173–200

Hanakahi LA, Dempsey LA, Li M-J, Maizels N (1997) Nucleolin is one component of the B cell-specific switch region binding protein, LR1. Proc Nat Acad Sci USA 94:3605–3610

Harriman GR, Bradley A, Das S, Rogers-Fani P, Davis AC (1996) IgA class switch in Iα exon deficient mice. Role of germline transcription in class switch recombination. J Clin Invest 97:477–485

Hasbold J, Lyons AB, Kehry MR, Hodgkin PD (1998) Cell division number regulates IgG1 and IgE switching of B cells following stimulation by CD40 ligand and IL-4. Eur J Immunol 28:1040–1051

Heldin CH, Miyazone K, Tendijke P (1997) TGF-β signalling from cell membrane to nucleus through the SMAD proteins. Nature 390:465–471

Hjulstrom S, Landin A, Holmdahl LJ, Heyman B (1995) No role of interleukin-4 in CD23/IgE-mediated enhancement of the murine antibody response in vivo. Eur J Immunol 25:1469–1472

Hodgkin PD, Lee J-H, Lyons AB (1996) B cell differentiation and isotype switching is related to division cycle number. J Exp Med 184:277–281

Hou J, Schindler U, Henzel WJ, Ho TC, Brasseur M, McKnight SL (1994) An interleukin-4-induced transcription factor: IL-4 Stat. Science 265:1701–1706

Huang S, Hendriks W, Althage A, Hemmi S, Bluethmann H, Kamijo R, Vilcek J, Zinkernagel RM, Aguet M (1993) Immune response in mice that lack the interferon-γ receptor. Science 259: 1742–1745

Hug BA, Wesselschmidt RL, Fiering S, Bender MA, Epner E, Groudine M, Ley TJ (1996) Analysis of mice containing a targeted deletion of the β-globin locus control region 5′ hypersensitive site 3. Mol Cell Biol 16:2906–2912

Hummel M, Kaminshka J, Dunnick W (1987) Switch region content of hybridomas: the two spleen cell IgH loci tend to rearrange to the same isotype. J Immunol 138:3539–3548

Huo L, Rothstein TL (1995) Receptor specific induction of individual AP-1 components in B lymphocytes. J Immunol 154:3300–3310

Ichiki T, Takahashi W, Watanabe T (1993) Regulation of the expression of human Cε germline transcript: Identification of a novel IL-4 responsive element. J Immunol 150:5408–5417

Iciek LA, Delphin SA, Stavnezer J (1997) CD40 cross-linking induces Igε germline transcripts in B cells via activation with NF-κB: Synergy with IL-4 induction. J Immunol 158:4769–4779

Irsch J, Hendriks R, Tesch H, Schuurman R, Radbruch A (1993) Evidence for a human IgG1 class switch program. Eur J Immunol 23:481–486

Irsch J, Irlenbusch S, Radl J, Burrows PD, Cooper MD, Radbruch AH (1994) Switch recombination in normal IgA1+ B lymphocytes. Proc Nat Acad Sci USA 91:1323–1327
Isakson PC, Pure E, Vitetta ES, Krammer PH (1982) T cell derived B cell differentiation factor(s). Effect on the isotype switch of murine B cells. J Exp Med 155:734–748
Islam KB, Nilsson L, Sideras P, Hammarström L, Smith CIE (1991) TGFβ1 induces germ-line transcripts of both IgA subclasses in human B lymphocytes. Int Immunol 3:1099–1106
Iwasato T, Shimizu A, Honjo T, Yamagishi H (1990) Circular DNA is excised by immunoglobulin class switch recombination. Cell 62:143–149
Jabara HH, Fu SM, Geha RS, Vercelli D (1990) CD40 and IgE: Synergism between anti-CD40 monoclonal antibody and interleukin 4 in the induction of IgE synthesis by highly purified human B cells. J Exp Med 172:1861–1864
Jacob J, Kassir R, Kelsoe G (1991) In situ studies of the primary immune response to (4-hydroxy-3-nitrophenyl)acetyl. I. The architecture and dynamics of the responding cell populations. J Exp Med 173:1165–1175
Jumper MD, Nishioka Y, Davis LS, Lipsky PE, Meek K (1995) Regulation of human B cell function by recombinant CD40 ligand and other TNF-related ligands. J Immunol 155:2369–2378
Jumper MD, Splawski JB, Lipsky PE, Meek K (1994) Ligation of CD40 induces sterile transcripts of multiple H chain isotypes in human B cells. J Immunol 152:438–454
Jung S, Rajewsky K, Radbruch A (1993) Shutdown of class switch recombination by deletion of a switch region control element. Science 259:984–987
Jung S, Siebenkotten G, Radbruch A (1994) Frequency of immunoglobulin E class switching is autonomously determined and independent of prior switching to other classes. J Exp Med 179:2023–2026
Kaplan MH, Schindler U, Smiley ST, Grusby MJ (1996) Stat6 is required for mediating responses to IL-4 and for the development of Th2 cells. Immunity 4:313–319
Kawabe T, Naka T, Yoshida K, Tanaka T, Fujiwara H, Suematsu S, Yoshida N, Kishimoto T, Kikutani H (1994) The immune responses in CD40-deficient mice: impaired immunoglobulin class switching and germinal center formation. Immunity 1:167–178
Kearney JF, Cooper MD, Lawson AR (1976) B cell differentiation induced by lipopolysaccharide. IV. Development of immunoglobulin class restriction in precursors of IgG synthesizing cells. J Immunol 117:1567–1572
Kearney JF, Lawton AR (1975) B lymphocyte differentiation induced by lipopolysaccharide. I. Generation of cells synthesizing four major immunoglobulin classes. J Immunol 115:671–676
Kenter AL, Birstein BK (1981) Chi, a promoter of generalized recombination in λ phage, is present in immunoglobulin genes. Nature 293:402–404
Kenter AL, Watson JV (1987) Cell cycle kinetics model of LPS-stimulated spleen cells correlates switch region rearrangements with S phase. J Immunol Methods 97:111–117
Kenter AL, Wuerffel R, Sen R, Jamieson CE, Merkulov GV (1993) Switch recombination breakpoints occur at nonrandom positions in the Sγ tandem repeat. J Immunol 151:4718–4731
Kepron MR, Chen Y-W, Uhr JW, Vitetta E (1989) IL-4 induces the specific rearrangement of γ1 genes on the expressed and unexpressed chromosomes of lipopolysaccharide activated normal murine B cells. J Immunol 143:334–339
Kim P-H, Kagnoff MF (1990) Transforming growth factor β1 increases IgA isotype switching at the clonal level. J Immunol 145:3773–3778
Kitani A, Strober W (1993) Regulation of Cγ subclass germ-line transcripts in human peripheral blood B cells. J Immunol 151:3478–3488
Kitani A, Strober W (1994) Differential regulation of Cα1 and Cα2 germ-line and mature mRNA transcripts in human peripheral blood B cells. J Immunol 153:1466–1477
Kluin PM, Kayano H, Zani VJ, Kluin-Nelemans HC, Tucker PW, Satterwhite E, Dyer MJS (1996) IgD class switching: identification of a novel recombination site in neoplastic and normal B cells. Europ J Immunol 25:3504–3508
Knapp MR, Liu C-P, Newell N, Ward RB, Tucker PW, Strober S, Blattner F (1982) Simultaneous expression of immunoglobulin μ and δ heavy chains by a cloned B-cell lymphoma: a single copy of the VH gene is shared by two adjacent CH genes. Proc Nat Acad Sci USA 79:2996–3000
Kohler I, Rieber EP (1993) Allergy-associated Iε and Fcε receptor II (CD23b) genes activated via binding of an interleukin-4-induced transcription factor to a novel responsive element. Eur J Immunol 23:3066–3071
Köntgen F, Grumont RJ, Strasser A, Metcalf D, Li R, Tarlinton D, Gerondakis S (1995) Mice lacking the c-rel proto-oncogene exhibit defects in lymphocyte proliferation, humoral immunity, and interleukin-2 expression. Genes Dev 9:1965–1977

Kopf M, LeGros G, Bachmann M, Lamers MC, Bluethmann H, Kohler G (1993) Disruption of the murine IL-4 gene blocks Th2 cytokine responses. Nature 362:245–248

Körthauer U, Graf D, Mages HW, Briere F, Padayachee M, Malcolm S, Ugazio AG, Notarangelo LD, Levinsky RJ, Kroczek RA (1993) Defective expression of T-cell CD40 ligand causes X-linked immunodeficiency with hyper-IgM. Nature 361:539–541

Kozmik Z, Wang S, Dorfler P, Adams B, Busslinger M (1992) The promoter of the CD19 gene is a target for the B-cell-specific transcription factor BSAP. Mol Cell Biol 12:2662–2672

Kraal G, Weissman IL, Butcher EC (1982) Germinal center cells: antigen specificity and changes in heavy chain class expression. Nature 298:377–379

Kühn R, Rajewsky K, Müller W (1991) Generation and analysis of interleukin-4 deficient mice. Science 254:707–710

Lalmanach-Girard A-C, Chiles TC, Parker DC, Rothstein TL (1993) T cell-dependent induction of NF-κB in B cells. J Exp Med 177:1215–1219

Lane P, Brocker T, Hubele S, Padovan E, Lanzavecchia A, McConnell F (1993) Soluble CD40 ligand can replace the normal T cell-derived CD40 ligand signal to B cells in T cell-dependent activation. J Exp Med 177:1209–1213

Lane P, Traunecker A, Hubele S, Inui S, Lanzavecchia A, Gray D (1992) Activated human T cells express a ligand for the human B cell-associated antigen CD40 which participates in T cell-dependent activation of B lymphocytes. Eur J Immunol 22:2573–2580

Lansford R, Manis JP, Sonoda E, Rajewsky K, Alt FW (1998) Ig heavy chain class switching in Rag-deficient mice. Int Immunol 10:325–332

Layton JE, Vitetta ES, Uhr JW, Krammer PH (1984) Clonal analysis of B cells induced to secrete IgG by T cell-derived lymphokines. J Exp Med 160:1850–1863

Lebman D, Coffman R (1988) Interleukin 4 causes isotype switching to IgE in T cell-stimulated clonal B cell cultures. J Exp Med 168:853–862

Lebman DA, Lee FD, Coffman RL (1990) Mechanism for transforming growth factor β and Il-2 enhancement of IgA expression in lipopolysaccharide-stimulated B cell cultures. J Immunol 144: 952–959

Lee SY, Lee SY, Choi Y (1997) TRAF-interacting protein (TRIP): a novel component of the tumor necrosis factor receptor (TNFR)- and CD30-TRAF signaling complexes that inhibits TRAF2-mediated NF-κB activtion. J Exp Med 185:1275–1285

Lehman CW, Carroll D (1991) Homologous recombination catalyzed by a nuclear extract from Xenopus oocytes. Proc Nat Acad Sci USA 88:10840–10844

Lehman CW, Trautman JK, Carroll D (1994) Illegitimate recombination in Xenopus: characterization of end-joined junctions. Nucl Acids Res 22:434–442

Lewis SM (1994) The mechanism of V(D)J joining: lessons from molecular, immunological, and comparative analyses. Adv Immunol 56:27–150

Li M-J, Maizels N (1997) Nuclear Rad51 foci induced by DNA damage are distinct from Rad51 foci associated with B cell activation and recombination. Exp Cell Res 237:93–100

Li M-J, Peakman M-C, Golub EI, Reddy G, Ward DC, Radding CM, Maizels N (1996) Rad51 expression and localization in B cells carrying out class switch recombination. Proc Nat Acad Sci USA 93:10222–01227

Liang F, Romanienko PJ, Weaver DT, Jeggo PA, Jasin J (1996) Chromosomal double-strand break repair in Ku80-deficient cells. Proc Nat Acad Sci USA 93:8929–8933

Liao F, Birshtein B, Busslinger M, Rothman P (1994) The transcription factor BSAP (NF-HB) is essential for immunoglobulin germ-line ε transcription. J Immunol 152:2904–2911

Liao F, Giannini SL, Birshtein B (1992) A nuclear DNA-binding protein expressed during early stages of B cell differentiation interacts with diverse segments within and 3′ of the Ig H chain gene cluster. J Immunol 148:2909–2917

Lim DS, Hasty P (1996) A mutation in mouse rad51 results in an early embryonic lethal that is suppressed by a mutation in p53. Mol Cell Biol 12:7133–7143

Lin S-C, Wortis HH, Stavnezer J (1998) The ability of CD40L, but not LPS, to initiate immunoglobulin switching to IgG1 is explained by differential induction of NF-κB/Rel proteins. Mol Cell Biol 18:5523–5532

Lin SC, Stavnezer J (1996) Activation of NF-κB/Rel by CD40 engagement induces the mouse germline immunoglobulin Cγ1 promoter. Mol Cell Biol 16:4591 4603

Lin Y-CA, Stavnezer J (1992) Regulation of transcription of the germline Igα constant region gene by an ATF element and by novel transforming growth factor-β1-responsive elements. J Immunol 149: 2914–2925

Linehan LA, Warren WD, Thompson PA, Grusby MJ, Berton MT (1998) STAT6 is required for IL-4-induced germline Ig gene transcription and switch recombination. J Immunol 161:302–310
Liu Y-J, deBouteiller O, Arpin C, Briere F, Galibert L, Ho S, Martinez-Valdez H, Banchereau J, Lebeque S (1996a) Normal human $IgD^{+}IgM^{-}$ germinal ceneter B cells can express up to 80 mutations in the variable region of their IgD transcripts. Immunity 4:603–613
Liu YJ, Malisan F, deBouteiller O, Guret C, Lebecque S, Banchereau J, Mills FC, Max EE, Martinez-Valdez H (1996b) Within germinal centers, isotype switching of immunoglobulin genes occurs after the onset of somatic mutation. Immunity 4:241–250
Lorenz M, Jung S, Radbruch A (1995) Switch transcripts in immunoglobulin class switching. Science 267:1825–1828
Lundgren M, Larrson C, Femino A, Xu M, Stavnezer J, Severinson E (1994) Activation of the Ig germline γ1 promoter: involvement of C/EBP transcription factors and their possible interaction with a NF-IL4 site. J Immunol 153:2983–2995
Lundgren M, Persson U, Larsson P, Magnusson C, Smith C, Hammerstrom L, Severinson E (1989) Interleukin 4 induces synthesis of IgE and IgG4 in human B cells. Eur J Immunol 19:1311–1315
Lundgren M, Strom L, Bergqvist LO, Skog S, Heiden T, Stavnezer J, Severinson E (1995) Cell cycle regulation of germline immunoglobulin transcription: potential role of Ets family members. Eur J Immunol 25:2042–2051
Lutzker S, Rothman P, Pollock R, Coffman R, Alt FW (1988) Mitogen- and IL-4-regulated expression of germline Igγ2b transcripts: Evidence for directed heavy chain class switching. Cell 53:177–184
Lycke N, Severinson E, Strober W (1990) Cholera toxin acts synergistically with IL-4 to promote IgG1 switch differentiation. J Immunol 145:3316–3324
Ma L, Hu B, Kenter AL (1997) Ig Sγ-specific DNA binding protein SNAP is related to the helix-loop-helix transcription factor E47. Intern Immunol 9:1021–1029
Madisen L, Groudine M (1994) Identification of a locus control region in the immunoglobulin heavy-chain locus that deregulates c-myc expression in plasmacytoma and Burkitt's lymphoma cells. Genes Dev 8:2212–2226
Maki R, Roder W, Traunecker A, Sidman C, Wabl M, Raschke W, Tonegawa S (1981) The role of DNA rearrangement and alternative RNA processing in the expression of IgD genes. Cell 24: 353–365
Malisan F, Briere F, Bridon J-M, Harindranath N, Mills FC, Max EE, Banchereau J, Martinez-Valdez H (1996) Interleukin-10 induces IgG isotype switch recombination in human CD40-activated naive B lymphocytes. J Exp Med 183:937–947
Mandler R, Chu CC, Paul W, Max EE, Snapper CM (1993a) Interleukin 5 induces Sμ-Sγ1 DNA rearrangement in B cells activated with dextran-anti-IgD antibodies and interleukin 4: a three component model for Ig class switching. J Exp Med 178:1577–1586
Mandler R, Finkelman FD, Levine AD, Snapper CM (1993b) IL-4 induction of IgE class switching by lipopolysaccharide-activated murine B cells occurs predominantly through sequential switching. J Immunol 150:407–418
Manis JP, Gu Y, Lansford R, Sonoda E, Ferrini R, Davidson L, Rajewsky K, Alt FW (1998a) Ku70 is required for late B cell development and immunoglobulin heavy chain switching J Exp Med 187: 2081–2089
Manis JP, van der Stoep N, Tian M, Ferrini R, Davidson L, Bottaro A, Alt FW (1998b) Class switching in B cells lacking 3′ Ig heavy chain enhancers. J Exp Med 188:1421–1431
Marcu KB, Lang RB, Stanton LW, Harris LJ (1982) A model for the molecular requirements of immunoglobulin heavy chain class switching. Nature 298:87–89
Massague J, Hata A, Liu F (1997) TGF-β signaling through the Smad pathway. Trends Cell Biol 7: 187–192
Matsuoka M, Yoshida K, Maeda T, Usuda S, Sakano H (1990) Switch circular DNA is formed in cytokine-treated mouse splenocytes: Evidence for intramolecular DNA deletion in immunoglobulin class switching. Cell 62:135–142
McHeyzer-Williams MG, McLean MJ, Lalor PA, Nossal GJV (1993) Antigen-driven B cell differentiation in vivo. J Exp Med 178:295–307
McIntyre TM, Kehry MR, Snapper CM (1995) Novel in vitro model for high-rate IgA class switching. J Immunol 154:3156–3161
McIntyre TM, Klinman DR, Rothman P, Lugo M, Dasch JR, Mond JJ, Snapper CM (1993) Transforming growth factor-β1 selectively stimulates immunoglobulin G2b secretion by lipopolysaccharide-activated murine B cells. J Exp Med 177:1031–1037

Messner B, Stutz AM, Albrecht B, Peiritsch S, Woisetschlager M (1997) Cooperation of binding sites for STAT6 and NFκB/rel in the IL-4-induced up-regulation of the human IgE germline promoter. J Immunol 159:3330–3337
Meyers S, Hiebert SW (1995) Indirect and direct disruption of transcriptional regulation in cancer: E2F and AML-1. Crit Rev Eukaryot Gene Expr 5:365–383
Mikita T, Campbell D, Wu P, Williamson K, Schindler U (1996) Requirements for interleukin-4-induced gene expression and functional characterization of Stat6. Mol Cell Biol 16:5811–5820
Mikita T, Kurama M, Shindler U (1998) Synergistic activation of the germline ε promoter mediated by Stat6 and C/EBPβ. J Immunol 161:1822–1828
Moore K W, Rogers J, Hunkapiller T, Early P, Nottenburg C, Weissman I, Bazin H, Wall R, Hood LE (1981) Expression of IgD may use both DNA rearrangement and RNA splicing mechanisms. Proc Nat Acad Sci USA 78:1800–1804
Morawetz RA, Gabriele L, Rizzo LVN, Noben-Trauth Kuhn R, Rajewsky K, Muller W, Doherty TM, Finkelman F, Coffman RL, Morse HC III (1996) IL-4 independent immunoglobulin class switch to IgE in the mouse. J Exp Med 184:1651–1661
Murre C, Baltimore D (1992) The helix-loop-helix motif: structure and function. In: Transcriptional regulation. Cold Spring Harbor Press, Plainview NY, pp 861–879
Müller JR, Giese T, Henry DL, Mushinski JF, Marcu KB (1998) Generation of switch hybrid DNA between Ig heavy chain-μ and downstream switch reigons in B lymphocytes. J Immunol 161: 1354–1362
Nakamura M, Kondo S, Sugai M, Nazarea M, Imamura S, Honjo T (1996) High frequency class switching of an $IgM^+$ B lymphoma clone CH12F3 to $IgA^+$ cells. Intl Immunol 8:193–201
Neurath MF, Max EE, Strober W (1995) Pax5 (BSAP) regulates the murine immunoglobulin 3′α enhancer by suppressing binding of NF-αP, a protein that controls heavy chain transcription. Proc Nat Acad Sci USA 92:5336–5340
Neurath MF, Strober W, Wakatsuki Y (1994) The murine Ig 3′α enhancer is a target site with repressor function for the B cell lineage-specific transcription factor BSAP (NF-HB, Sα-BP). J Immunol 153:730–742
Nilsson L, Sideras P (1993) The human $I_\alpha 1$ and $I_\alpha 2$ germline promoter elements: proximal positive and distal negative elements may regulate the tissue specific expression of Cα1 and Cα2 germline transcripts. Int Immunol 5:271–282
Nitschke L, Kosco MH, Kohler G, Lamers MC (1993) IgD-deficient mice can mount normal immune responses to thymus-independent and -dependent antigens. Proc Nat Acad Sci USA 90:1887–1891
Noben-Trauth N, Kropf P, Muller I (1996) Susceptibility to leishmania major infection in interleukin-4-deficient mice. Science 271:987–990
Noelle RJ, Roy M, Shepherd DM, Stamenkovic I, Ledbetter JA, Aruffo A (1992a) A 39-kDa protein on activated helper T cells binds CD40 and transduces the signal for cognate activation of B cells. Proc Nat Acad Sci USA 89:6550–6554
Noelle RJ, Shepherd DM, Fell HP (1992b) Cognate interaction between T helper cells and B cells VII. Role of contact and lymphokines in the expression of germ-line and mature γ1 transcripts. J Immunol 149:1164–1169
O'Rourke L, Tooze R, Fearon DT (1997) Co-receptors of B lymphocytes. Curr Opin Immunol 9:324–329
O'Shea JJ (1997) Review: jaks, STATs, cytokine signal transduction and immune regulation: are we there yet? Immunity 7:1–12
Owens JD, Finkelman FD, Mountz JD, Mushinski JF (1991) Nonhomologous recombination at sites within mouse $J_H$ -Cδ locus accompanies Cμ deletion and switch to IgD secretion. Mol Cell Biol 11:5660–5670
Peakman MC, Maizels N (1998) Localization of splenic B cells activated for switch recombination by in situ hybridization with Igγ1 switch transcript and Rad51 probes. J Immunol 161:4008–4015
Punnonen J, Aversa G, Cocks BG, McKenzie ANJ, Menon S, Zurawski G, de Waalmalefyt R, Devries JE (1993) Interleukin 13 induces interleukin 4-independent IgG4 and IgE synthesis and CD23 expression by human B cells. Proc Nat Acad Sci USA 90:3730–3734
Purkerson JM, Isakson PC (1992) Interleukin 5 (IL-5) provides a signal that is required in addition to IL-4 for isotype switching to IgG1 and IgE. J Exp Med 175:973–982
Purkerson JM, Newberg M, Wise G, Lynch KR, Isakson PC (1988) Interleukin 5 and interleukin 2 cooperate with interleukin 4 to induce IgG1 secretion from anti-Ig-treated B cells. J Exp Med 168:1175–1180

Qiu G, Harriman GR, Stavnezer J (1999) Iα exon-replacement mice synthesize a spliced HPRT-Cα transcript which may explain their ability to switch to IgA: Inhibition of switching to IgG in these mice. Int Immunol 11:37–46
Qiu G, Stavnezer J (1998) Over-expression of BSAP/Pax-5 inhibits switching to IgA and enhances switching to IgE in the I29μ B cell line. J Immunol 161:2906–2918
Rathmell WK, Chu G (1994a) A DNA end-binding factor involved in double-strand break repair and V(D)J recombination. Mol Cell Biol 14:4741–4748
Rathmell WK, Chu G (1994b) Involvement of the Ku autoantigen in the cellular response to DNA double strand breaks. Proc Nat Acad Sci USA 91:7623–7627
Reichel M, Nelson BH, Greenberg PD, Rothman PB (1997) The IL-4 receptor α-chain cytoplasmic domain is sufficient for activation of JAK-1 and STAT6 and the induction of IL-4 specific gene expression. J Immunol 158:5860–5867
Renshaw BRWCF III, Armitage RJ, Campbell KA, Liggitt D, Wright B, Davison BL, Maliszewski CR (1994) Humoral immune response in CD40 ligand-deficient mice. J Exp Med 180:1889–1900
Roes J, Rajewsky K (1993) Immunoglobulin D-deficient mice reveal an auxiliary receptor function for IgD in antigen-mediated recruitment of B cells. J Exp Med 177:45–55
Rolink A, Melchers F, Andersson J (1996) The SCID but not the RAG-2 gene product is required for Sμ-Sε heavy chain class switching. Immunity 5:319–330
Roper RL, Brown DM, Phipps RP (1995) Prostaglandin E2 promotes B lymphocyte Ig isotype switching to IgE. J Immunol 154:162–170
Roper RL, Conrad DH, Brown DM, Warner GL, Phipps RP (1990) Prostaglandin E2 promotes IL-4-induced IgE and IgG1 synthesis. J Immunol 145:2644–2651
Roth DB, Wilson JH (1986) Nonhomologous recombination in mammalian cells: Role for short sequence homologies in the joining reaction. Mol Cell Biol 6:4295–4304
Rothman P, Li SC, Gorham B, Glimcher L, Alt FW, Boothby M (1991) Identification of a conserved LPS/IL-4 responsive element located at the promoter of germline ε transcripts. Mol Cell Biol 11: 5551–5561
Rothman P, Lutzker S, Cook W, Coffman R, Alt FW (1988) Mitogen plus interleukin 4 induction of Cε transcripts in B lymphoid cells. J Exp Med 168:2385–2389
Rothman P, Lutzker S, Gorham B, Stewart V, Coffman R, Alt FW (1990) Structure and expression of germline immunoglobulin γ3 heavy chain gene transcripts: implications for mitogen and lymphokine directed class-switching. Int Imunol 2:621–627
Rousset F, Garcia E, Banchereau J (1991) Cytokine-induced proliferation and immunoglobulin production of human B lymphocytes triggered through their CD40 antigen. J Exp Med 173:705–710
Savelkoul HFJ, Lebman DA, Benner R, Coffman R (1988) Increase of precursor frequency and clonal size of murine IgE secreting cells by IL-4. J Immunol 141:749–755
Schindler C, Kashleva H, Pernis A, Pine R, Rothman P (1994) STF-IL-4: a novel IL-4-induced signal transducing factor. EMBO J 13:1350–1356
Schindler U, Wu P, Roth M, Brasseur M, McKnight S (1995) Components of a Stat recognition code: Evidence for two layers of molecular selectivity. Immunity 2:687–697
Schmitz J, Radbruch A (1989) An interleukin 4-induced DNase I hypersensitive site indicates opening of the γ1 switch region prior to switch recombination. Int Immunol 1:570–575
Schultz CL, Rothman P, Kuhn R, Kehry M, Muller W, Rajewsky K, Alt FW, Coffman RL (1992) T helper cell membranes promote IL-4-independent expression of germ-line Cγ1 transcripts in B cells. J Immunol 149:60–64
Screpanti I, Romani L, Musiani P, Modesti A, Fattori E, Lazzaro D, Sellitto C, Scarpa S, Bellavia D, Lattanzio G, et al. (1995) Lymphoproliferative disorder and imbalanced T-helper response in C/EBP beta-deficient mice. EMBO J 14:1932–41
Seidl K, Manis JP, Bottaro A, Zhang J, Davidson L, Kisselgof A, Oettgen H, Alt FW (1999) Position-dependent inhibition of class switch recombination by PGK-NEO$^r$ cassetts inserted into the Ig heavy chain constant region locus. Proc Nat Acad Sci USA 96:3000–3005
Seidl KJ, Bottaro A, Vo A, Zhang J, Davidson L, Alt FW (1998) An expressed neo$^r$ cassette provides required functions of the Iγ2b exon for class switching. Int Immunol 10:101–110
Severinson E, Bergstedt-Lindqvist S, Loo W, Vanderloo, Fernandez C (1982) Characterization of the IgG response induced by polyclonal B cell activators. Immunol Rev 67:73–85
Severinson E, Fernandez C, Stavnezer J (1990) Induction of germ-line immunoglobulin heavy chain transcripts by mitogens and interleukins prior to switch recombination. Eur J Immunol 20:1079–1084
Severinson-Gronowicz E, Doss C, Schroder J (1979) Activation to IgG secretion by lipopolysaccharide requires several proliferation cycles. J Immunol 123:2057–2062

Sha W C, Liou H-C, Tuomanen EI, Baltimore D (1995) Targeted disruption of the p50 subunit of NF-κB leads to multifocal defects in immune responses. Cell 80:321–330

Shapira SK, Vercelli D, Jabara HH, Fu SM, Geha RS (1992) Molecular analysis of the induction of immunoglobulin E synthesis in human B cells by interleukin 4 and engagement of CD40 antigen. J Exp Med 175:289–292

Shen C-H, Stavnezer J (1998) Interaction of Stat6 and NF-κB: Direct association and synergistic activation of IL-4 induced transcription. Mol Cell Biol 18:3395–3404

Shi MJ, Stavnezer J (1998) CBFa3 (AML2) is induced by TGF-β1 to bind and activate the mouse germline Ig α promoter. J Immunol 161:6751–6760

Shimoda K, Deursen JV, Sangster MY, Sarawar SR, Carson RT, Tripp RA, Chu C, Quelle FW, Nosaka T, Vignali DAA, Doherty PC, Grosveld G, Paul WE, Ihle JN (1996) Lack of IL-4 induced Th2 response and IgE class switching in mice with disrupted Stat6 gene. Nature 380:630–633

Shockett P, Stavnezer J (1991) Effect of cytokines on switching to IgA and α germline transcripts in the B lymphoma I29μ: transforming growth factor-ß activates transcription of the unrearranged Cα gene. J Immunol 147:4374–4383

Shockett P, Stavnezer J (1993) Inhibitors of poly(ADP-ribose) polymerase increase antibody class switching. J Immunol 151:6962–6976

Shparago N, Zelazowski P, Jim L, McIntyre TM, Stuber E, Pecanha LMT, Kehry MR, Mond JJ, Max EE, Snapper CM (1996) Interleukin-10 selectively regulates murine isotype switching. Intl Immunol 8:781–790

Sideras P, Mizuta T-R, Kanamori H, Suzuki N, Okamoto M, Kuze K, Ohno H, Doi S, Fukuhara S, Hassan MS, Hammarstrom L, Smith E, Shimizu A, Honjo T (1989) Production of sterile transcripts of Cγ genes in an IgM-producing human neoplastic B cell line that switches to IgG-producing cells. Int Immunol 1:631–642

Siebenkotten G, Esser C, Wabl M, Radbruch A (1992) The murine IgG1/IgE class switch program. Eur J Immunol 22:1827–1834

Singh M, Birshtein BK (1993) NF-HB (BSAP) is a repressor of the murine immunoglobulin heavy-chain 3′α enhancer at early stages of B-cell differentiation. Mol Cell Biol 13:3611–3622

Singh M, Birshtein BK (1996) Concerted repression of an immunoglobulin heavy-chain enhancer, 3′ αE(hs1,2). Proc Nat Acad Sci USA 93:5336–5340

Smith CA, Gruss HJ, Davis T, Anderson D, Farrah T, Baker D, Sutherland GR, Brannan CI, Copeland NG, Jenkins NA, Grabstein KH, Gliniak B, McAlister IB, Fanslow W, Alderson M, Falk B, Gimpel S, Gillis S, Din WS, Goodwin RG, Armitage RJ (1993) CD30 antigen, a marker for Hodgkin's lymphoma, is a receptor whose ligand defines an emerging family of cytokines with homology to TNF. Cell 73:1349–1360

Snapper CM, Finkelman F (1998) Immunoglobulin class switching. In: Fundamental immunology, 4th edn. Raven Press, New York, pp 831–861

Snapper CM, Mond JJ (1993) Towards a comprehensive view of Ig class switching. Imm Today 14:15–17

Snapper CM, Mond JJ (1996) Commentary: A model for induction of T cell-independent humoral immunity is response to polysaccharide antigens. J Immunol 157:2229–2233

Snapper CM, Paul WE (1987) Interferon-γ and B cell stimulatory factor-1 reciprocally regulate Ig isotype production. Science 236:944–947

Snapper CM, Pecanha LMT, Levine AD, Mond JJ (1991) IgE class switching is critically dependent upon the nature of the B cell activator, in addition to the presence of IL-4. J Immunol 147:1163–1170

Snapper CM, McIntyre TM, Mandler R, Pecanha LMT, Finkelman FD, Lees A, Mond JJ (1992) Induction of IgG3 secretion by interferon γ: a model for T cell-independent class switching in response to T cell-independent type 2 antigens. J Exp Med 175:1367–1371

Snapper CM, Kehry MR, Castle BE, Mond JJ (1995) Multivalent, but not divalent, antigen receptor cross-linkers synergize with CD40 ligand for induction of Ig synthesis and class switching in normal murine B cells. J Immunol 154:1177–1187

Snapper CM, Rosasa FR, Zelazowski P, Moorman MA, Kehry MR, Bravo R, Weih F (1996a) B cells lacking RelB are defective in proliferative responses, but undergo normal B cell maturation to Ig secretion and Ig class switching. J Exp Med 184:1537–1541

Snapper CM, Zelazowski P, Rosas FR, Kehry MR, Tian M, Baltimore D, Sha WC (1996b) B cells from p50/NF-κB knockout mice have selective defects in proliferation, differentiation, germline $C_H$ transcription and Ig class switching. J Immunol 156:183–191

Sonoda E, Matsumoto R, Hitoshi Y, Ishii T, Sugimoto M, Araki S, Tominaga A, Yamaguchi N, Takatsu K (1989) Transforming growth factor β induces IgA production and acts additively with interleukin 5 for IgA production. J Exp Med 170:1415–1420

Speck NA, Terryl S (1995) A new transcription factor family associated with human leukemias. Crit Rev Eukaryot Gene Expr 5:337–364
Splawski JB, Fu SM, Lipsky PE (1993) Immunoregulatoy role of CD40 in human B cell differentiation. J Immunol 150:1276–1285
Stavnezer J (1995a) Regulation of antibody production and class switching by TGFβ. J Immunol 155:1647–1651
Stavnezer J (1995b) TGF-β regulation of B cell proliferation, differentiation, and class switching. In: Regulation of immunoglobulin synthesis and class switching. Wiley & Sons Ltd, Sussex, pp 289–324
Stavnezer J (1996a) Antibody class switching. Adv Immunol 61:79–146
Stavnezer J (1996b) Immunoglobulin class switching. Curr Opin Immunol 8:199–205
Stavnezer J, Abbott J, Sirlin S (1984) Immunoglobulin heavy chain switching in cultured I.29 murine B lymphoma cells: commitment to an IgA or IgE switch. Curr Top Microbiol Immunol 113:109–116
Stavnezer J, Radcliffe G, Lin Y-C, Nieutupski J, Berggren L, Sitia R, Severinson E (1988) Immunoglobulin heavy-chain switching may be directed by prior induction of transcripts from constant-region genes. Proc Natl Acad Sci USA 85:7704–7708
Stavnezer-Nordgren J, Sirlin S (1986) Specificity of immunoglobulin heavy chain switch correlates with activity of germline heavy chain genes prior to switching. EMBO J 5:95–102
Taccioli GE, Rathbun G, Olz E, Stamato T, Jeggo PA, Alt FW (1993) Impairment of V(D)J recombination in double-strand break repair mutants. Science 260:207–210
Takeda K, Tanaka T, Shi W, Matsumoto M, Minami M, Kashiwamura S-I, Nakanishi K, Yoshida N, Kishimoto T, Akira S (1996) Essential role of Stat6 in IL-4 signalling. Nature 380:627–630
Teale JM, Klinman NR (1984) Control of the production of different classes of antibody. In: Fundamental Immunology. Raven Press, New York, pp 519–535
Thienes CP, de Monte L, Monticelli S, Busslinger M, Gould HJ, Vercelli D (1997) The transcription factor B cell-specific activator protein (BSAP) enhances both IL-4- and CD40-mediated activation of the human epsilon germline promoter. J Immunol 158:5874–82
Thornton BP, Vetcicka V, Ross GD (1994) Natural antibody and complement-mediated antigen processing and presentation. J Immunol 152:1727–1737
Toellner K-M, Gulbranson-Judge A, Taylor DR, Sze DM-Y, MacLennan ICM (1996) Immunoglobulin switch transcript production in vivo related to the site and time of antigen-specific B cell activation. J Exp Med 183:2303–2312
Toellner K-M, Luther SA, Sze DM-Y, Choy RK-W, Taylor DR, MacLennan ICM, Acha-Orbea H (1998) Th1 and Th2 characteristics start to develop during T cell priming and are associated with an immediate ability to induce immunoglobulin class switching. J Exp Med 187:1193–1204
Turaga PSD, Berton MT, Teale JM (1993) Frequency of B cells expressing germ-line $\gamma_1$ transcripts upon IL-4 induction. J Immunol 151:1383–1390
Urbanek P, Wang Z-Q, Fetka I, Wagner EF, Busslinger M (1994) Complete block of early B cell differentiation and altered patterning of the posterior midbrain in mice lacking Pax5/BSAP. Cell 79:901–912
Vajdy M, Kosco-Vilbois MH, Kopf M, Kohler G, Lycke N (1995) Impaired mucosal immune responses in IL-4 targeted mice. J Exp Med 181:41–53
van Kooten C, Banchereau J (1996) CD40-CD40 ligand: A multifunctional receptor-ligand pair. Adv Immunol 61:1–77
Vercelli D, Jabara HH, Arai K-I, Geha RS (1989) Induction of human IgE synthesis requires interleukin 4 and T/B interactions involving the T cell receptor/CD3 complex and MHC class II antigens. J Exp Med 169:1295–1307
Vitetta E, Ohara J, Meyers C, Layton J, Krammer P, Paul W (1985) Serological, biochemical and functional identity of B cell stimulatory factor 1 and B cell differentiation factor for IgG1. J Exp Med 162:1726–1731
von Schwedler U, Jack HM, Wabl M (1990) Circular DNA is a product of the immunoglobulin class switch rearrangement. Nature 345:452–456
Vonderweid T, Kopf M, Kohler G, Langhorne J (1994) The immune response to Plasmodium chabaudi malaria in interleukin-4-deficient mice. Eur J Immunol 24:2285–2293
Wabl M, Meyer J, Beck-Engeser G, Tenkhoff M, Burrows PD (1985) Critical test of a sister chromatid exchange model for the immunoglobulin heavy-chain class switch. Nature 313:687–689
Wakatsuki Y, Neurath MF, Max EE, Strober W (1994) The B cell-specific transcription factor BSAP regulates B cell proliferation. J Exp Med 179:1099–1108
Wakatsuki Y, Strober W (1993) Effect of downregulation of germline transcripts on immunoglobulin A isotype differentiation. J Exp Med 178:129–138

Wallin JJ, Gackstetter ER, Koshland ME (1998) Dependence of BSAP repressor and activator functions on BSAP concentration. Science 279:1961–1964

Wang Z-Q, Stingl C, Morrison M, Jantsch M, Los K Schulze-Osthoff, Wagner EF (1997) PARP is important for genomic stability but dispensable in apoptosis. Genes Dev 11:2347–2358

Warren WD, Berton MT (1995) Induction of germline γ1 and εIg gene expression in murine B cells: interleukin 4 and the CD40 ligand-CD40 interaction provide distinct but synergistic signals. J Immunol 155:5637–5646

Waters SH, Saikh KU, Stavnezer J (1989) A B-cell-specific nuclear protein that binds to DNA sites 5′ to immunoglobulin Sα tandem repeats is regulated during differentiation. Mol Cell Biol 9:5594–5601

Weih F, Warr G, Yang H, Bravo R (1997) Multifocal defects in immune responses in RelB-deficient mice. J Immunol 158:5211–5218

Weinstein PD, Cebra JJ (1991) The preference for switching to IgA expression by Peyers patch germinal center B cells is likely due to the intrinsic influence of their microenvironment. J Immunol 147: 4126–4135

White MB, Word CJ, Humphries CG, Blattner FR, Tucker PW (1990) Immunoglobulin D switching can occur through homologous recombination in human B cells. Mol Cell Biol 10:3690–3699

Whitmore AC, Prowse DM, Haughton G, Arnold LW (1991) Ig isotype switching in B lymphocytes. The effect of T cell-derived interleukins, cytokines, cholera toxin, and antigen on isotype switch frequency of a cloned B cell lymphoma. Int Immunol 3:95–103

Williams M, Hanakahi LA, Maizels N (1993) Purification and properties of LR1, an inducible DNA binding protein from mammalian B lymphocytes. J Biol Chem 268:13731–13737

Williams M, Maizels N (1991) LR1, a lipopolysaccharide-responsive factor with binding sites in the immunoglobulin switch regions and heavy-chain enhancer. Genes Dev 5:2353–2361

Winter E, Krawinkel U, Radbruch A (1987) Directed Ig class switch recombination in activated murine B cells. EMBO J 6:1663–1671

Wuerffel R, Jamieson CE, Morgan L, Merkulov GV, Sen R, Kenter AL (1992) Switch recombination breakpoints are strictly correlated with DNA recognition motifs for immunoglobulin Sγ3 DNA-binding proteins. J Exp Med 176:339–349

Wuerffel RA, Du J, Thompson RJ, Kenter AL (1997) Ig Sγ3 DNA-specific double strand breaks are induced in mitogen-activated B cells and are implicated in switch recombination. J Immunol 159:4139–4144

Xie X-Q, Pardali E, Holm M, Sideras P, Grundstrom T. AML and Ets proteins regulate the Iα1 germ line promoter. Eur J Immunol 29:488–498

Xu J, Foy TM, Laman JD, Elliott EA, Dunn JJ, Waldschmidt TJ, Elsemore J, Noelle RJ, Flavell RA (1994) Mice deficient for the CD40 ligand. Immunity 1:423–432

Xu L, Kim MG, Marcu KB (1992) Properties of B cell stage specific and ubiquitous factors binding to immunoglobulin heavy chain gene switch regions. Int Immunol 4:875–887

Xu L, Rothman P (1994) IFN-γ represses ε germline transcription and subsequently down-regulates switch recombination to ε. Int Immunol 6:515–521

Xu M, Stavnezer J (1992) Regulation of transcription of immunoglobulin germ-line γ1 RNA: analysis of the promoter/enhancer. EMBO J 11:145–155

Yancopolous G, de Pihno R, Zimmerman K, Lutzker S, Rosenberg N, Alt FW (1986) Secondary rearrangement events in pre B cells: $V_HDJ_H$ replacement by LINE-1 sequence and directed class switching. EMBO J 5:3259–3266

Yasui H, Akahor Y, Hiran M, Yamada K, Kurosawa Y (1989) Class switch from μ to δ is mediated by homologous recombination between σμ and Σψ sequences in human immunoglobulin gene loci. Eur J Immunol 19:1399–1403

Yuan D (1986) Regulation of IgD synthesis in murine neonatal B lymphocytes. Mol Cell Biol 6: 1015–1022

Yuan D, Tucker PW (1984) The transcriptional regulation of the μ-δ heavy chain locus in normal murine B lymphocytes. J Exp Med 160:564–583

Yuan D, Witte PL (1988) Transcriptional regulation of μ and δ gene expression in bone marrow pre-B and B lymphocytes. J Immunol 140:2808–2814

Zawel L, Dai JL, Buckhaults P, Zhou S, Kinzler KW, Vogelstein B, Kern SE (1998) Human Smad3 and Smad4 are sequence-specific transcription activators. Mol Cell 1:611–617

Zelazowski P, Carrasco D, Rosas FR, Moorman MA, Bravo R, Snapper CM (1997a) B cells genetically deficient in the c-Rel transactivation domain have selective defects in germline $C_H$ transcription and Ig class switching. J Immunol 159:3133–3139

Zelazowski P, Collins JT, Dunnick W, Snapper CM (1995) Antigen receptor cross-linking differentially regulates germ-line $C_H$ ribonucleic acid expression in murine B cells. J Immunol 154: 1223–1231

Zelazowski P, Max EE, Kehry MR, Snapper CM (1997b) Regulation of Ku expression in normal murine B cells by stimuli that promote switch recombination. J Immunol 159:2559–2562

Zhang J, Bottaro A, Li S, Stewart V, Alt FW (1993) A selective defect in IgG2b switching as a result of targeted mutation of the Iγ2b promoter and exon. EMBO J 12:3529–3537

Zhang Y, Cao HJ, Graf B, Meekins H, Smith TJ, Phipps RP (1998) Cutting edge: CD40 engagement up-regulates cyclooxygenase-2 expression and prostaglandin $E_2$ production in human lung fibroblasts. J Immunol 160:1053–1057

Zurawski G, Vries JED (1994) Interleukin 13, an interleukin 4-like cytokine that acts on monocytes and B cells, but not on T cells. Immunol Today 15:19–26

# Subject Index

# Current Topics in Microbiology and Immunology

Volumes published since 1989 (and still available)

Vol. 204: **Saedler, Heinz; Gierl, Alfons (Eds.):** Transposable Elements. 1995. 42 figs. IX, 234 pp. ISBN 3-540-59342-X

Vol. 205: **Littman, Dan R. (Ed.):** The CD4 Molecule. 1995. 29 figs. XIII, 182 pp. ISBN 3-540-59344-6

Vol. 206: **Chisari, Francis V.; Oldstone, Michael B. A. (Eds.):** Transgenic Models of Human Viral and Immunological Disease. 1995. 53 figs. XI, 345 pp. ISBN 3-540-59341-1

Vol. 207: **Prusiner, Stanley B. (Ed.):** Prions Prions Prions. 1995. 42 figs. VII, 163 pp. ISBN 3-540-59343-8

Vol. 208: **Farnham, Peggy J. (Ed.):** Transcriptional Control of Cell Growth. 1995. 17 figs. IX, 141 pp. ISBN 3-540-60113-9

Vol. 209: **Miller, Virginia L. (Ed.):** Bacterial Invasiveness. 1996. 16 figs. IX, 115 pp. ISBN 3-540-60065-5

Vol. 210: **Potter, Michael; Rose, Noel R. (Eds.):** Immunology of Silicones. 1996. 136 figs. XX, 430 pp. ISBN 3-540-60272-0

Vol. 211: **Wolff, Linda; Perkins, Archibald S. (Eds.):** Molecular Aspects of Myeloid Stem Cell Development. 1996. 98 figs. XIV, 298 pp. ISBN 3-540-60414-6

Vol. 212: **Vainio, Olli; Imhof, Beat A. (Eds.):** Immunology and Developmental Biology of the Chicken. 1996. 43 figs. IX, 281 pp. ISBN 3-540-60585-1

Vol. 213/I: **Günthert, Ursula; Birchmeier, Walter (Eds.):** Attempts to Understand Metastasis Formation I. 1996. 35 figs. XV, 293 pp. ISBN 3-540-60680-7

Vol. 213/II: **Günthert, Ursula; Birchmeier, Walter (Eds.):** Attempts to Understand Metastasis Formation II. 1996. 33 figs. XV, 288 pp. ISBN 3-540-60681-5

Vol. 213/III: **Günthert, Ursula; Schlag, Peter M.; Birchmeier, Walter (Eds.):** Attempts to Understand Metastasis Formation III. 1996. 14 figs. XV, 262 pp. ISBN 3-540-60682-3

Vol. 214: **Kräusslich, Hans-Georg (Ed.):** Morphogenesis and Maturation of Retroviruses. 1996. 34 figs. XI, 344 pp. ISBN 3-540-60928-8

Vol. 215: **Shinnick, Thomas M. (Ed.):** Tuberculosis. 1996. 46 figs. XI, 307 pp. ISBN 3-540-60985-7

Vol. 216: **Rietschel, Ernst Th.; Wagner, Hermann (Eds.):** Pathology of Septic Shock. 1996. 34 figs. X, 321 pp. ISBN 3-540-61026-X

Vol. 217: **Jessberger, Rolf; Lieber, Michael R. (Eds.):** Molecular Analysis of DNA Rearrangements in the Immune System. 1996. 43 figs. IX, 224 pp. ISBN 3-540-61037-5

Vol. 218: **Berns, Kenneth I.; Giraud, Catherine (Eds.):** Adeno-Associated Virus (AAV) Vectors in Gene Therapy. 1996. 38 figs. IX,173 pp. ISBN 3-540-61076-6

Vol. 219: **Gross, Uwe (Ed.):** Toxoplasma gondii. 1996. 31 figs. XI, 274 pp. ISBN 3-540-61300-5

Vol. 220: **Rauscher, Frank J. III; Vogt, Peter K. (Eds.):** Chromosomal Translocations and Oncogenic Transcription Factors. 1997. 28 figs. XI, 166 pp. ISBN 3-540-61402-8

Vol. 221: **Kastan, Michael B. (Ed.):** Genetic Instability and Tumorigenesis. 1997. 12 figs.VII, 180 pp. ISBN 3-540-61518-0

Vol. 222: **Olding, Lars B. (Ed.):** Reproductive Immunology. 1997. 17 figs. XII, 219 pp. ISBN 3-540-61888-0

Vol. 223: **Tracy, S.; Chapman, N. M.; Mahy, B. W. J. (Eds.):** The Coxsackie B Viruses. 1997. 37 figs. VIII, 336 pp. ISBN 3-540-62390-6

Vol. 224: **Potter, Michael; Melchers, Fritz (Eds.):** C-Myc in B-Cell Neoplasia. 1997. 94 figs. XII, 291 pp.
ISBN 3-540-62892-4

Vol. 225: **Vogt, Peter K.; Mahan, Michael J. (Eds.):** Bacterial Infection: Close Encounters at the Host Pathogen Interface. 1998. 15 figs. IX, 169 pp.
ISBN 3-540-63260-3

Vol. 226: **Koprowski, Hilary; Weiner, David B. (Eds.):** DNA Vaccination/Genetic Vaccination. 1998. 31 figs. XVIII, 198 pp.
ISBN 3-540-63392-8

Vol. 227: **Vogt, Peter K.; Reed, Steven I. (Eds.):** Cyclin Dependent Kinase (CDK) Inhibitors. 1998. 15 figs. XII, 169 pp.
ISBN 3-540-63429-0

Vol. 228: **Pawson, Anthony I. (Ed.):** Protein Modules in Signal Transduction. 1998. 42 figs. IX, 368 pp. ISBN 3-540-63396-0

Vol. 229: **Kelsoe, Garnett; Flajnik, Martin (Eds.):** Somatic Diversification of Immune Responses. 1998. 38 figs. IX, 221 pp.
ISBN 3-540-63608-0

Vol. 230: **Kärre, Klas; Colonna, Marco (Eds.):** Specificity, Function, and Development of NK Cells. 1998. 22 figs. IX, 248 pp.
ISBN 3-540-63941-1

Vol. 231: **Holzmann, Bernhard; Wagner, Hermann (Eds.):** Leukocyte Integrins in the Immune System and Malignant Disease. 1998. 40 figs. XIII, 189 pp.
ISBN 3-540-63609-9

Vol. 232: **Whitton, J. Lindsay (Ed.):** Antigen Presentation. 1998. 11 figs. IX, 244 pp.
ISBN 3-540-63813-X

Vol. 233/I: **Tyler, Kenneth L.; Oldstone, Michael B. A. (Eds.):** Reoviruses I. 1998. 29 figs. XVIII, 223 pp.
ISBN 3-540-63946-2

Vol. 233/II: **Tyler, Kenneth L.; Oldstone, Michael B. A. (Eds.):** Reoviruses II. 1998. 45 figs. XVI, 187 pp.
ISBN 3-540-63947-0

Vol. 234: **Frankel, Arthur E. (Ed.):** Clinical Applications of Immunotoxins. 1999. 16 figs. IX, 122 pp. ISBN 3-540-64097-5

Vol. 235: **Klenk, Hans-Dieter (Ed.):** Marburg and Ebola Viruses. 1999. 34 figs. XI, 225 pp. ISBN 3-540-64729-5

Vol. 236: **Kraehenbuhl, Jean-Pierre; Neutra, Marian R. (Eds.):** Defense of Mucosal Surfaces: Pathogenesis, Immunity and Vaccines. 1999. 30 figs. IX, 296 pp.
ISBN 3-540-64730-9

Vol. 237: **Claesson-Welsh, Lena (Ed.):** Vascular Growth Factors and Angiogenesis. 1999. 36 figs. X, 189 pp.
ISBN 3-540-64731-7

Vol. 238: **Coffman, Robert L.; Romagnani, Sergio (Eds.):** Redirection of Th1 and Th2 Responses. 1999. 6 figs. IX, 148 pp.
ISBN 3-540-65048-2

Vol. 239: **Vogt, Peter K.; Jackson, Andrew O. (Eds.):** Satellites and Defective Viral RNAs. 1999. 39 figs. XVI, 179 pp.
ISBN 3-540-65049-0

Vol. 240: **Hammond, John; McGarvey, Peter; Yusibov, Vidadi (Eds.):** Plant Biotechnology. 1999. 12 figs. XII, 196 pp.
ISBN 3-540-65104-7

Vol. 241: **Westblom, Tore U.; Czinn, Steven J.; Nedrud, John G. (Eds.):** Gastroduodenal Disease and Helicobacter pylori. 1999. 35 figs. XI, 313 pp.
ISBN 3-540-65084-9

Vol. 242: **Hagedorn, Curt H.; Rice, Charles M. (Eds.):** The Hepatitis C Viruses. 2000. 47 figs. IX, 379 pp. ISBN 3-540-65358-9

Vol. 243: **Famulok, Michael; Winnacker, Ernst-L.; Wong, Chi-Huey (Eds.):** Combinatorial Chemistry in Biology. 1999. 48 figs. IX, 189 pp. ISBN 3-540-65704-5

Vol. 244: **Daëron, Marc; Vivier, Eric (Eds.):** Immunoreceptor Tyrosine-Based Inhibition Motifs. 1999. 20 figs. VIII, 179 pp.
ISBN 3-540-65789-4

Vol. 245/I: **Justement, Louis B.; Siminovitch, Katherine A. (Eds.):** Signal Transduction and the Coordination of B Lymphocyte Development and Function I. 2000. 22 figs. XVI, 274 pp. ISBN 3-540-66002-X

MIX
Papier aus verantwortungsvollen Quellen
Paper from responsible sources
FSC® C105338

If you have any concerns about our products,
you can contact us on
**ProductSafety@springernature.com**

In case Publisher is established outside the EU,
the EU authorized representative is:
**Springer Nature Customer Service Center GmbH**
**Europaplatz 3, 69115 Heidelberg, Germany**

Printed by Libri Plureos GmbH
in Hamburg, Germany